# Laser Technology for Materials Processing

# Laser Technology for Materials Processing

Guest Editor

**Nuno Ferreira**

MDPI

Basel • Beijing • Wuhan • Barcelona • Belgrade • Novi Sad • Cluj • Manchester

*Guest Editor*
Nuno Ferreira
i3N and Department of Physics
University of Aveiro
Aveiro
Portugal

*Editorial Office*
MDPI AG
Grosspeteranlage 5
4052 Basel, Switzerland

This is a reprint of the Special Issue, published open access by the journal *Materials* (ISSN 1996-1944), freely accessible at: https://www.mdpi.com/journal/materials/special_issues/8L360G4P0Q.

For citation purposes, cite each article independently as indicated on the article page online and as indicated below:

Lastname, A.A.; Lastname, B.B. Article Title. *Journal Name* **Year**, *Volume Number*, Page Range.

**ISBN 978-3-7258-8189-5 (Hbk)**
**ISBN 978-3-7258-8190-1 (PDF)**
**https://doi.org/10.3390/books978-3-7258-8190-1**

# Contents

# About the Editor

**Nuno Ferreira**

Nuno Ferreira is an Assistant Professor at the Department of Physics of the University of Aveiro (UA) and a Senior Researcher at **i3N** (Institute of Nanostructures, Nanomodelling and Nanofabrication). With a PhD in Physical Engineering from the University of Aveiro, he has built a distinguished career at the intersection of Fundamental Physics and Advanced Materials Engineering. His core research expertise lies in **laser-based materials processing**, with a particular focus on the **Laser Floating Zone (LFZ)** technique. He has been a key member of the **PAMD** (Physics of Advanced Materials and Devices) group, where he leads investigations into the synthesis and optimization of functional materials, including **high-temperature thermoelectric materials** for sustainable energy harvesting.

Beyond material synthesis, he is actively involved in the digital transformation of manufacturing processes. His research focuses on enhancing the precision and efficiency of laser processing and surface engineering to develop high-performance materials. This interdisciplinary approach aims to bridge the gap between laboratory-scale physics and industrial applications, particularly in the fields of green energy and biomaterials.

With a prolific scientific output of over *100 peer-reviewed publications* and a strong track record in student supervision and project management, he has been a committed advocate for the role of Physics in driving industrial innovation. He is currently focused on advancing the synergy between experimental physics and the paradigms of **Industry 4.0/5.0**, contributing to a more resilient and sustainable technological future.

# Preface

This Reprint, "Laser Technology for Materials Processing", explores the subject of laser–matter interactions related to their properties and industrial applications. Its scope includes surface modification, ablation, and sintering. The aim is to present sustainable, high-precision methods for material functionalization. Motivated by the need for eco-friendly manufacturing, this scientific work addresses challenges in process efficiency. It is addressed to researchers and engineers in materials science and physics.

**Nuno Ferreira**
*Guest Editor*

materials

MDPI

*icle*

# Iicrostructural Investigation of Process Parameters Dedicated to aser Powder Bed Fusion of AlSi7Mg0.6 Alloy

usz Kluczyński [1,*], Tomáš Dražan [2], Zdeněk Joska [2], Jakub Łuszczek [1], Robert Kosturek [1]
d Katarzyna Jasik [1]

[1] Institute of Robots & Machine Design, Faculty of Mechanical Engineering, Military University of Technology, Gen. S. Kaliskiego St., 00-908 Warsaw, Poland; jakub.luszczek@wat.edu.pl (J.Ł.); robert.kosturek@wat.edu.pl (R.K.); katarzyna.jasik@wat.edu.pl (K.J.)

[2] Department of Mechanical Engineering, Faculty of Military Technology, University of Defence, 662 10 Brno, Czech Republic; tomas.drazan@unob.cz (T.D.); zdenek.joska@unob.cz (Z.J.)

* Correspondence: janusz.kluczynski@wat.edu.pl; Tel.: +48-261-837-646

**Abstract:** This study presents a microstructural investigation of the printing parameters of an AlSi7Mg0.6 alloy produced by powder bed fusion (PBF) using laser beam melting (LB/M) technology. The investigation focused on the effects of laser power, exposure velocity, and hatching distance on the microhardness, porosity, and microstructure of the produced alloy. The microstructure was characterized in the plane of printing on a confocal microscope. The results showed that the printing parameters significantly affected the microstructure, whereas the energy density had a major effect. Decreasing the laser power and decreasing the hatching distance resulted in increased porosity and the increased participation of non-melted particles. A mathematical model was created to determine the porosity of a 3D-printed material based on three printing parameters. Microhardness was not affected by the printing parameters. The statistical model created based on the porosity investigation allowed for the illustration of the technological window and showed certain ranges of parameter values at which the porosity of the produced samples was at a possible low level.

**Keywords:** additive manufacturing; metallography; porosity; microhardness; AlSi7Mg0.6

## 1. Introduction

Additive manufacturing (AM) has become an increasingly popular method that continues to transform the manufacturing industry [1–3]. Today's additive technologies have advanced capabilities regarding material options, speed, precision, cost-effectiveness, and the possibility of producing complex geometries with high accuracy [4]. AM shows promise for the production of components with alloys that cannot be processed with traditional processes and also unique microstructure [5,6]. Each of the dozens of additive technologies has a large number of possibilities but also has limitations, and therefore, it is necessary to select the appropriate additive technology to obtain the desired product properties [7]. Some of the common problems faced by most additive technologies are porosity, hot cracking, the anisotropy of the mechanical properties, and surface quality. The investigation of mechanical properties and microstructural investigation is essential to optimize the printing parameters and achieve high-quality components [2,6–13]. Powder bed fusion-selective laser melting (PBF-LB/M) is one of the most widely used additive manufacturing technologies, where metal powder is melted layer-by-layer using a laser to create the desired component [2,9–11,14–16]. In recent years, there has been a growing interest in the use of PBF-LB/M to produce components for aerospace, automotive, biomedical applications, and a lot of other branches [4,5,12,16,17]. PBF produces a specific microstructure by melting the material and making a melting pool in the printing direction. Heat directed to the melting point is distributed through the build material by thermal conduction to form an oriented fine-grained microstructure [7,17]. Therefore, microhardness measurements are often used to correlate the microstructure with the mechanical properties of the printed components [14]. The parts produced by SEM typically require post-processing to achieve the desired surface finish and mechanical properties [18]. Post-processing can include removing support structures, sandblasting or polishing, and heat treatment to reduce residual stress [14]. Of course, the post-processing steps are tailored to the specific material and application requirements. Aluminum alloys, such as AlSi7Mg0.6, are widely used materials in these industries due to their strength-to-weight ratio and corrosion resistance [14]. However, the quality of the final product is highly dependent not only on powder quality [19] and heat treatment but especially on printing parameters, including laser power, hatching distance, and exposure velocity [7,17]. Optimizing these printing parameters for this material is challenging because each parameter affects a lot of variables [1,13]. The idea of creating a mathematical model showing the theoretical porosity using these printing parameters seems to be very advantageous. Aluminum materials are known for their high thermal conductivity and, thus, the occurrence of residual stress and warping [1,9,20]. Today's aluminum alloy research deals with printed PBF-LB/M technologies, mostly addressing quality depending on the printing orientation [4,8,11], research on material properties and production of metal powder [3,21], the formation of defects, corrosion, stress corrosion, and the effect of heat treatment [4,5,10,18,20,22] or by modeling and predicting the properties of printed parts [17]. In the case of research on the effect of changing the printing parameters, the articles bend in the range of different values of printing parameters than in our research [1,6] or dealing with other specifications [12,13,15,16,22,23]. Martucci et al. [1] conducted research on the printing strategies of the AlSi10Cu8Mg aluminum alloy using laser powers below 100 W and lower printing speeds. His results indicated an improvement in terms of porosity but difficulties with the problem of crack formation and delamination. By optimizing the process parameters, he was able to produce high-quality samples. Arvieu et al. [6] then dealt with the interpretation of the relative density of AlSi7Mg0.6 and AM205 aluminum alloys printed by PBF-LB/M.

The aluminum alloy sample investigated in our work consists of only five different printing processes with significantly different parameters. However, the results of his research show that the energy density has a direct influence on the porosity of the printed sample, and it is possible to achieve relative densities even higher than the 100% reported in the literature. No research has been found that has attempted to develop a mathematical model of the printing parameters and their effect on the porosity of the AlSi7Mg0.6 alloy produced by PBF-LB/M technology. In the present state of the art, there is a small amount of research related to microstructural evaluation with this range of processing parameters of PBF-LB/M technology and implementing them to mathematical models. Therefore, understanding energy density combined with the dependent printing parameters and their influence on the porosity and microstructure of the printed components is crucial for designing and producing components with the best possible material and mechanical properties.

The aim of our research is to develop a comprehensive understanding of the relationship between energy density and various printing parameters in the production of aluminum alloy components using PBF-LB/M technology. Specifically, we aim to investigate how different combinations of printing parameters affect the porosity and microstructure of AlSi7Mg0.6 alloy samples. By analyzing the influence of energy density along with other printing parameters, we seek to develop a mathematical model that accurately predicts the porosity levels and microstructural characteristics of printed components. This research is essential for optimizing the manufacturing process and achieving components with superior material and mechanical properties.

## 2. Materials and Methods

The 10 mm cube-shaped samples were designed in Magic 19 software (Leuven, Belgium, Materialise), and the same software was used to provide printing parameters, such as laser power, layer thickness, exposure velocity, and hatching distance (clearance between laser irradiation lines). The number of geometries was added to the top of the cube for easy recognition of the samples and to avoid confusion between the samples. These cubes were spaced around the print bed so that they would not interfere with each other during the process, as shown in Figure 1. The printing of the samples was performed on an SLM 125HL 3D PBF-LB/M printer (SLM Solution, Lubeck, Germany). Before the process, the equipment was properly cleaned of previous processes and materials. Before loading and calibration, the material was dried in a SUSLAB-BIO-005 laboratory dryer (Adverti, Łódź, Poland) for 24 h at 90 °C temperature. An aluminum alloy AlSi7Mg0.6 (AC-42200), gas atomized powder supplied by SLM Solutions (SLM Solution, Lubeck, Germany) was used as the base material. The powder particles were characterized by a spherical particle size of 20–63 µm. The thermal conductivity of this material was 150–170 W/(m·K). The chemical composition of this material, according to the material data sheet, is given in Table 1 below. The quality of the metal powder was checked by a scanning electron microscope (SEM) Jeol JSM-6610 (Jeol, Tokyo, Japan).

**Figure 1.** Distribution of printed samples on the substrate plate.

**Table 1.** Chemical composition of AlSi7Mg0.6 in wt.% [24].

| Al | Cu | Fe | Mg | Nb+Ta | Mn | Si | Ti | N | Zn | Other Each | Another Total |
|---|---|---|---|---|---|---|---|---|---|---|---|
| Bal. | 0.05 | 0.19 | 0.45–0.70 | / | 0.1 | 6.50–7.50 | 0.25 | / | 0.07 | 0.03 | 0.10 |

The observed process parameters selected were laser power, exposure velocity, and the hatching distance. The thickness of the powder layer was fixed with a constant value of 0.03 mm. The parameter which is a function of these 3 parameters is the energy density. For this, Equation (1) was used.

$$E_V = \frac{P_L}{h \times v_s \times H} \quad (1)$$

$E_V$—energy per unit volume (J/mm$^3$);
$P_L$—laser power (W);
h—hatching distance (mm);
$v_s$—exposure velocity (mm/s);
H—thickness of the powder layer (mm).

Many combinations of these parameters were created for the mathematical model and are published in Table 2. In total, 37 test samples were printed. Three groups of samples were created according to the dominant parameters. The first set (1–13) were samples printed with high laser power (>300W) and high exposure velocity (>1300 mm/s), while the second set (14–25) were printed with laser power (<200W) and hatching distance (<0.1), and the third set (26–37) with laser power (<200W) and hatching distance (<0.1).

**Table 2.** Combination of printing parameters.

| Sample | Power P [W] | Layer Thickness [mm] | Exposure Velocity [mm/s] | Hatching Distance [mm] | Energy Density [J/mm³] |
|---|---|---|---|---|---|
| 1 | 350 | 0.03 | 1650 | 0.13 | 54.390 |
| 2 | 350 | 0.03 | 1650 | 0.14 | 50.505 |
| 3 | 350 | 0.03 | 1650 | 0.12 | 58.923 |
| 4 | 385 | 0.03 | 1650 | 0.13 | 59.829 |
| 5 | 315 | 0.03 | 1650 | 0.13 | 48.951 |
| 6 | 350 | 0.03 | 1815 | 0.13 | 49.446 |
| 7 | 350 | 0.03 | 1485 | 0.13 | 60.433 |
| 8 | 350 | 0.03 | 1650 | 0.15 | 47.138 |
| 9 | 350 | 0.03 | 1650 | 0.11 | 64.279 |
| 10 | 400 | 0.03 | 1650 | 0.13 | 62.160 |
| 11 | 280 | 0.03 | 1650 | 0.13 | 43.512 |
| 12 | 350 | 0.03 | 1980 | 0.13 | 45.325 |
| 13 | 350 | 0.03 | 1320 | 0.13 | 67.988 |
| 14 | 190 | 0.03 | 800 | 0.13 | 60.897 |
| 15 | 180 | 0.03 | 800 | 0.13 | 57.692 |
| 16 | 170 | 0.03 | 800 | 0.13 | 54.487 |
| 17 | 190 | 0.03 | 1000 | 0.12 | 52.778 |
| 18 | 180 | 0.03 | 1000 | 0.12 | 50.000 |
| 19 | 170 | 0.03 | 1000 | 0.12 | 47.222 |
| 20 | 190 | 0.03 | 1000 | 0.11 | 57.576 |
| 21 | 180 | 0.03 | 1000 | 0.11 | 54.545 |
| 22 | 170 | 0.03 | 1000 | 0.11 | 51.515 |
| 23 | 190 | 0.03 | 1200 | 0.11 | 47.980 |
| 24 | 180 | 0.03 | 1200 | 0.11 | 45.455 |
| 25 | 170 | 0.03 | 1200 | 0.11 | 42.929 |
| 26 | 190 | 0.03 | 1200 | 0.1 | 52.778 |
| 27 | 190 | 0.03 | 1200 | 0.09 | 58.642 |
| 28 | 190 | 0.03 | 1200 | 0.08 | 65.972 |
| 29 | 190 | 0.03 | 1200 | 0.07 | 75.397 |
| 30 | 190 | 0.03 | 1200 | 0.06 | 87.963 |
| 31 | 190 | 0.03 | 1200 | 0.05 | 105.556 |
| 32 | 190 | 0.03 | 1000 | 0.05 | 126.667 |
| 33 | 190 | 0.03 | 1000 | 0.06 | 105.556 |
| 34 | 190 | 0.03 | 1000 | 0.07 | 90.476 |
| 35 | 190 | 0.03 | 800 | 0.07 | 113.095 |
| 36 | 190 | 0.03 | 800 | 0.06 | 131.944 |
| 37 | 190 | 0.03 | 800 | 0.07 | 113.095 |

To avoid thermal influence during the removal of the samples from the printing substrate, wire electric discharge cutting technology was used on the Accutex AL 400SA machine (AccuteX Technologies Co., Ltd., Taichung City, Taiwan). For experiments and evaluation, samples were cut in the cross-section through the whole layers of each sample by means of a metallographic saw with direct cooling. Subsequently, the samples were mounted in resin and subjected to grounding and polishing. Porosity measurements were performed on all 37 polished samples using a Keyence (Osaka, Japan) VHX-7000 digital microscope. An analysis of the structural quality of the samples produced using different printing parameters was divided into two stages. In the first step, there was stitching of sequentially exposed areas, and then the ratio of the porous area detected through the optical device to the total area of the sample was determined. The measurements for all specimens were performed using the same settings related to the detection of defects. The porosity value presented in this manuscript is based on a single measurement for each plane. Afterward, the specimens were re-polished and etched for metallographic evaluation. Mounted and polished samples were etched with Keller's reagent used for aluminum alloys. It is a mixture of nitric acid ($HNO_3$), hydrochloric acid (HCl), and hydrofluoric acid (HF). The etching times were significantly different depending on the printing parameters, and thus, multiple sample preparations were necessary. Multiple images were taken of each sample at three magnifications for a complex evaluation. The final test was the measurement of microhardness on an automated Struers DuraScan hardness tester (Struers, Copenhagen, Denmark) using the Vickers method.

Due to the material properties, the measurements were made under a load of 100 g and evaluated with a microscope at 40x magnification. The measurements were made with 6 indentations at 1 mm from the edge and 4 from the individual indentations. Before each indent, the position was checked and corrected to avoid local porosity.

Due to the time-consuming nature of the parameter selection process, a decision was made to attempt the use of a Design of Experiment (DoE) analysis. The mathematical models from the DoE area, with their predictive capabilities for the dependent variable, have the potential to reduce the time required for parameter selection by minimizing the number of parameter groups that need to be tested. Consequently, it decreases the time intensity of such research endeavors. The authors of [25,26] adopted such an approach in the parameter selection process for manufacturing model elements using the PBF-LB/M technique. One of the types of models in this context is the quadratic surface regression model. This model amalgamates polynomial regression models with fractional models. It takes into account not only the influence of independent variables on the predictor but also the interactions among variables. The independent variables encompass the components of energy density derived from formula (1), except for the layer thickness, which remained a constant value. Meanwhile, the described value (dependent variable/predictor) is the porosity value. The general form of the model is presented in formula (2):

$$y = \beta_0 + \beta_1 x_1 + \beta_2 x_2 + \beta_3 x_3 + \beta_{11} x_1^2 + \beta_{22} x_2^2 + \beta_{33} x_3^2 + \beta_{12} x_1 x_2 + \beta_{13} x_1 x_3 + \beta_{23} x_2 x_3 + \in \quad (2)$$

In Equation (2), the individual components represent the following:

y—porosity;

$x_1$—laser power;

$x_2$—exposure velocity;

$x_3$—hatching distance;

$\beta_m$ and $\beta_{mn}$ (for m = 1,2,3; n = 1,2,3)—the regression coefficients for individual variables and their product combinations;

$\in$—modeling residual error.

The source data for the model comprises actual porosity measurements from samples produced using the groups listed in Table 2. However, for a better fit of the model to the source values (as described by the coefficient $R^2$), a portion of parameter groups was excluded, where the values of a given independent variable did not repeat in at least three other groups. The excluded groups were as follows: 2, 4–8, 10–13, 26–28, 31–32. The necessity of employing this approach arises from a significant increase in the model prediction error ($\in$) due to the utilization of individual parameter groups where the values of specific independent variables do not recur in other groups. Additionally, a *p*-value was determined. The estimation of the p coefficient involved created an analysis of variance (ANOVA) table, and its relevance pertained to conducting statistical tests. Each component's statistical significance was assessed using a threshold of $p < 0.05$. All calculations and charts were made using the Statistica software (TIBICO Software Inc., Palo Alto, CA, USA). The purpose of using this model is to determine the group of parameters according to which the porosity value of the model elements is close to 0%. The results of the statistical model serve as a source of groups of parameters that can be tested in further stages of work on the discussed material.

## 3. Results

Particle analysis of the powder used for printing the samples was performed on a Tescan Mira4 scanning electron microscope (Tescan Brno Czechia). Figure 2 shows a relatively uniform powder particle size distribution at lower resolution. The size of the particles varied from 2 to 30 μm, and they had a spherical or slightly ellipsoidal shape. In the detailed image (Figure 2—right side), a spherical particle with a diameter of 40 μm is shown with fused small satellites with a diameter of approximately 3 μm, which indicates that recycled powder was used for printing, where both new and used powders were used.

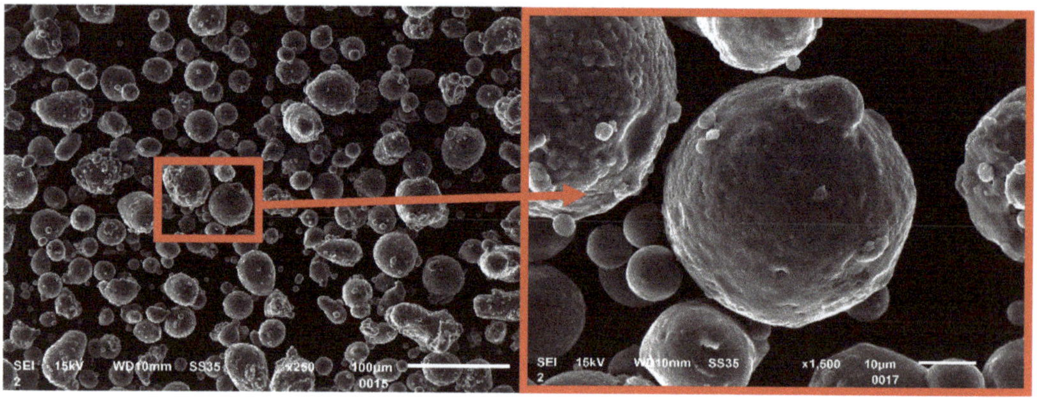


**Figure 2.** SEM image of the powder particles used to print the samples.

### 3.1. Porosity Measurement

A crucial factor in Metal Additive Manufacturing (MAM) methods that allow the quality verification of produced parts is porosity [26–31], which is why this parameter was used for the first step selection of the produced parts for further analysis. The percentage of porosity in all produced samples ranged from 0.38% to 1.71%. On average, the value of pore representation was 1%, with a standard deviation equal to 0.29%. The area of the measured pores was very small, and therefore, the observation of maximum and average perimeter sizes from all pores was chosen. The maximum measured circumference was 3.71 mm, and the smallest was 0.31 mm. The average perimeter for the samples was 0.1 mm, and the smallest average perimeter was 0.04 mm. Thus, the majority of pores had a perimeter limit close to 0.31 mm. Sample No. 22 was the one with the highest proportion of porosity, equal to 1.36% (shown in Figure 3).

A properly made drying process made the gas porosity (caused by moisture) marginal. Visible increased porosity on the side parts of the sample was caused by the same outline perimeter shell parameters used for all samples. For the measurement, these porosities were not taken into account. Only green-highlighted pores were considered in porosity calculations. As is visible in Figure 3, irregular voids are mostly caused by a lack of fusion phenomenon [32,33]. This factor is strictly dependent on process parameters, which is why the number of this kind of void fluctuation depends on the process parameters used.

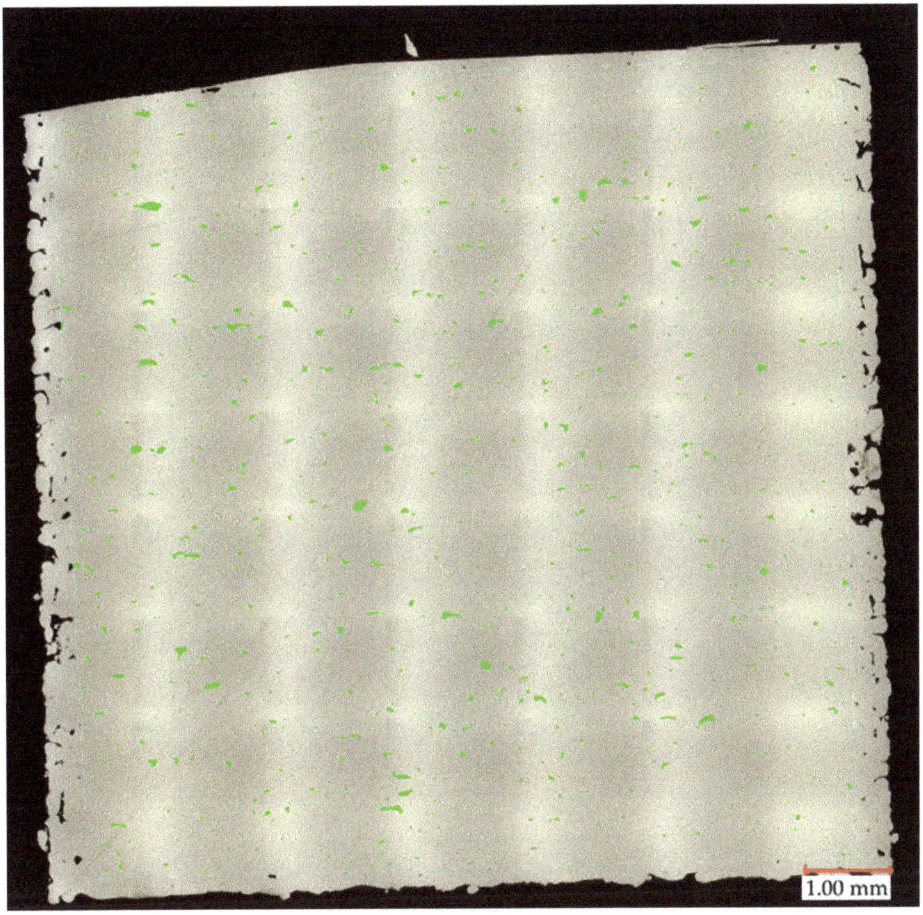


**Figure 3.** Merged images on areal porosity of sample 22 characterized by the highest porosity.

Due to the large volume of samples created and the large number of possible combinations of the influence of individual printing parameters, two groups of samples (Table 3) were selected for this article, which differed significantly in certain process parameters but were the same or similar in importance for others. The first group of samples that were selected for metallographic evaluation were samples 12 and 36, which had the lowest and highest energy density of the entire group of samples.

**Table 3.** Process parameters used for further microstructural analysis.

| Sample | Power [W] | Layer Thickness [mm] | Exposure Velocity [mm/s] | Hatching Distance [mm] | Energy Density [J/mm$^3$] | Measured Porosity (%) |
|---|---|---|---|---|---|---|
| 25 | 170 | 0.03 | 1200 | 0.11 | 42.929 | 1.67 |
| 36 | 190 | 0.03 | 800 | 0.06 | 131.944 | 1.53 |

In these samples, the laser power differed by only 20 W, with the main difference appearing in the different exposure velocities, which was one-third higher in the sample with lower energy density, and the hatching distance, which was almost twice as high in the sample with lower energy density. The microstructure of the No. 25 and No. 36 samples is shown in Figures 4 and 5.

**Figure 4.** Sample 25's microstructure at different magnifications.

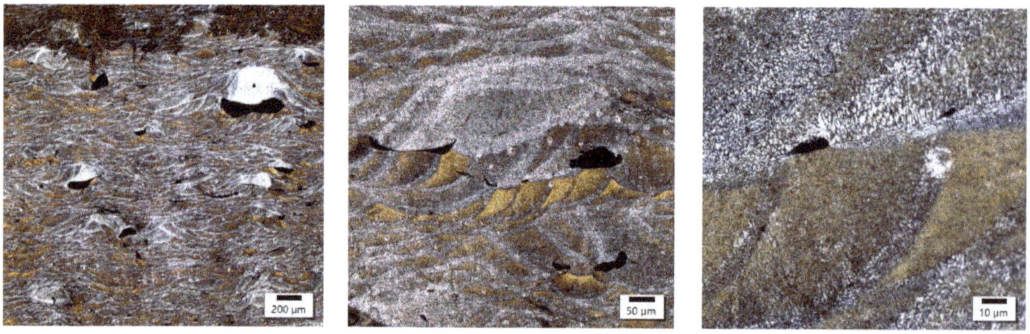


**Figure 5.** Sample 36's microstructure at different magnifications.

Both samples show a similar microstructure, which is typical for 3D printing with bands of melted pools of irregular shape. Despite some similarities at the macrostructural level, the sample produced at a higher energy density was characterized by a clearly smaller size of melted pools. The established size of the melted pools equaled 32.4 $\pm$ 7.3 µm and 23.1 $\pm$ 7.1 µm for the samples with a lower and higher energy density, respectively. These results indicate a reduction in the average size of about 30%. On the microstructural level, the lower energy density is reflected in the slightly finer columnar cellular substructure and the presence of ultrafine equiaxed columns with a width of about 8 µm and cell size below 1 µm. The increase in energy density causes the columnar substructure to coarse, and at the same time, no ultrafine equiaxed cells in the zone were found. In addition, in the sample with a higher energy density, the presence of microcracks oriented perpendicularly to the direction of columnar grain growth was reported, which suggests their solidification nature.

### *3.2. Metallography—Same Energy Density*

In the case of using the same energy density, samples number 2 and 18 (shown in Table 4) were chosen for further analysis, which have an energy density of 50 J/mm$^3$. Their selection was justified to analyze how different laser powers affect the microstructure. Such a comparison is very important from an economic point of view.

**Table 4.** Samples 2 and 18 and their printing parameters.

| Sample | Power [W] | Layer Thickness [mm] | Exposure Velocity [mm/s] | Hatching Distance [mm] | Energy Density [J/mm$^3$] | Measured Porosity (%) |
|---|---|---|---|---|---|---|
| 2 | 350 | 0.03 | 1650 | 0.14 | 50.505 | 0.78 |
| 18 | 180 | 0.03 | 1000 | 0.12 | 50.000 | 1.05 |

The etching of microstructure (Figures 6 and 7) by Keller's reagent for samples produced by laser power at around 350 W took twice the time compared to samples produced under laser power at around 180 W. Sample 2 came out on top with the power of 350 W while sample 18 exhibited a power of 180 W in magnification.

**Figure 6.** Sample 2's microstructure at different magnifications.

**Figure 7.** Sample 18's microstructure at different magnifications.

Comparing the samples obtained with various levels of laser power and constant energy density, the conclusion can be drawn that their macrostructures are very similar in terms of melted pool sizes. Despite the presence of porosity, observations did not reveal a noticeable number of other imperfections (e.g., microcracks). The microstructures of the investigated samples exhibited similar features, with the width of melted pool interfaces in both cases at about 10 µm wide. The only visible difference was the grainy structure of the interfaces and the fact that the sample with lower power (180 W) was characterized by finer grains in this area. The participation of equiaxed grains at the expense of columnar grains was also higher than for the sample obtained with 350 W. Referring these observations to the results for the samples obtained with different values of energy density, the conclusion can be drawn that the value of energy density had the highest impact on the macrostructural level (melted pool size, imperfections, melted pool interface width). At the same time, with an optimized energy density, the laser power can shape the microstructure of the AlSiMg alloy. Based on the performed observations, it can be stated that a lower value of laser power (180 W) is more profitable in terms of grain size in the melted pool interface for investigated energy density (about 50 $J/mm^3$).

The image from sample number 36 (Figure 8) shows the association of incompletely melted powder with a coarse-grained structure and columnar grains oriented in accordance with the heat flow direction. In the image of sample 29 (Figure 9), the boundary between the fine-grained and coarse-grained structures was separated by a crack. The columnar grains, oriented in accordance with the direction of heat flow, crystallized from the probable point of contact of the two interfaces.

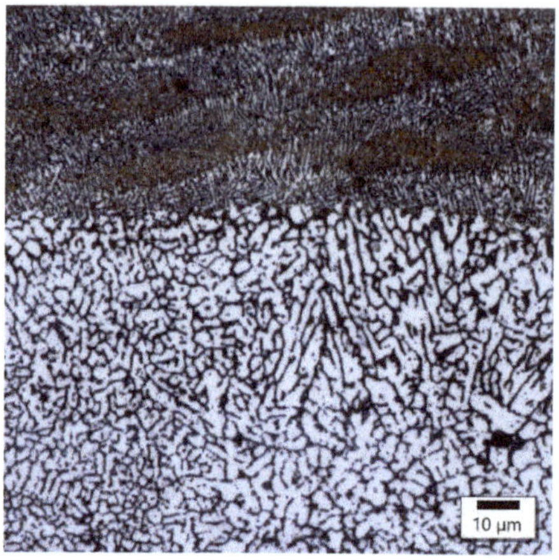


**Figure 8.** Sample 36's coarse-grained structure.

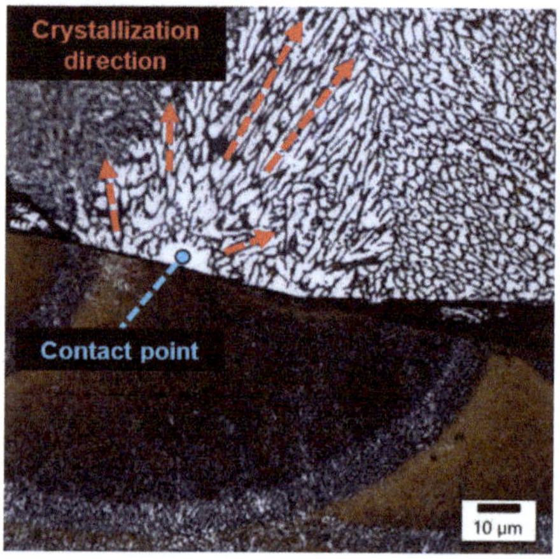


**Figure 9.** Sample 29's point of contact of imperfectly melted material.

In sample number 26 (Figure 10), two solidification cracks separated the fine-grained zone from the area of the columnar grain, whose growth was dictated by a small, 10 µm width bridge allowing for further heat flow. The image of sample 34 (Figure 11) then indicates that porosity formed at the boundary of the melting pools.

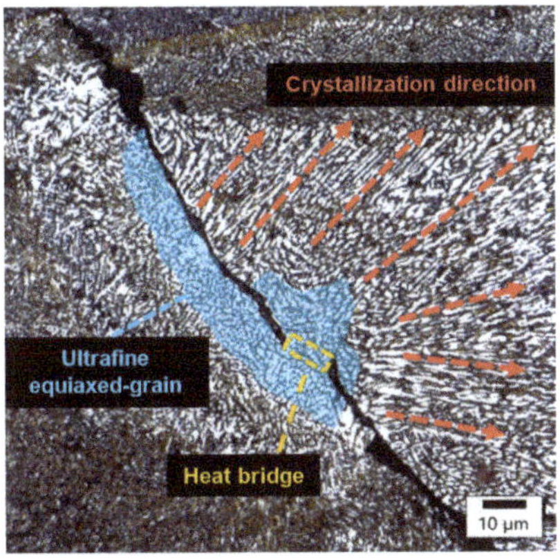


**Figure 10.** Sample 26 showing cracks in the structure due to heat bridge.

**Figure 11.** Sample 34 showing porosity on the boundaries of melting pools.

In sample 3 (Figures 12 and 13), microcracks initiated from a spherical pore formed by moisture in the matrix were observed. Similarly, a crack was observed in the middle of the melting pool. These are probably the thermal cracks that are commonly found at the boundaries of melting pools.

**Figure 12.** 3 Sample 3's spherical pore.

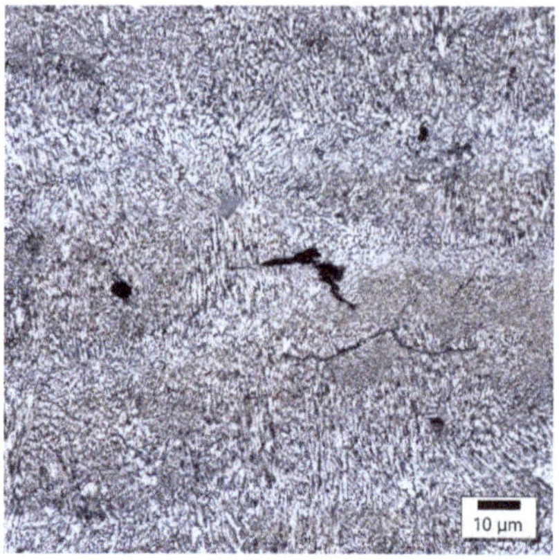


**Figure 13.** Sample 3 showing cracks in the middle melting pool.

### *3.3. Microhardness Measurement*

Hardness measurement results are presented on the chart in Figure 14. The average hardness value was 128 HV 0.1 with a standard deviation of 2.5 HV 0.1. The majority of results for individual parameter/sample groups fell within this range of deviation. Hence, it can be concluded that within the investigated range of parameter values, no significant changes in properties in relation to material hardness occurred. These findings are further supported by microstructural analysis results. The dotted red line indicates the mean value for all measurements.

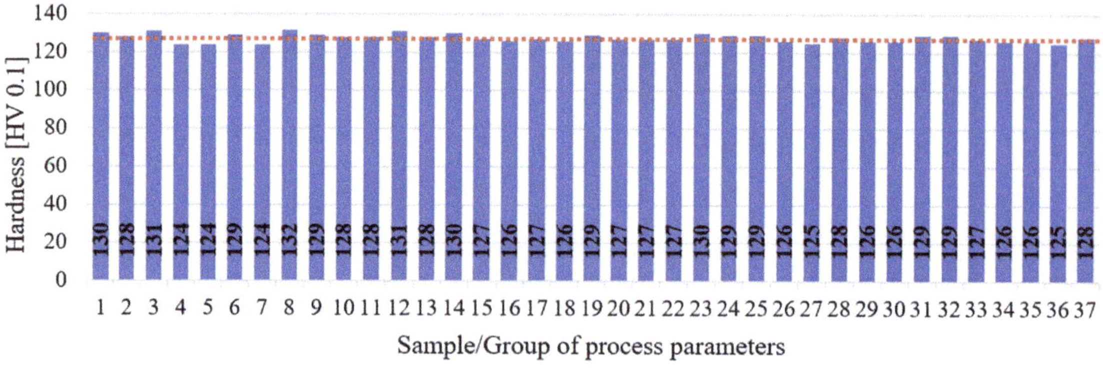


**Figure 14.** The results of measurement tests of AlSi7Mg0.6 samples produced by means of PBF-LB/M.

In reference to the literature, other researchers achieved hardness levels ranging from 45 to 120 HV, with power density values ranging between 10 and 242 J/mm$^3$ [14,34]. These hardness values were lower by approximately 8 HV compared to the average value obtained in this study and the maximum value found in the literature. However, the direct comparison of these values is challenging as numerous other additional factors, such as exposure strategy or layer height, which were not considered in this study, could significantly influence the final outcome. The settings utilized were default settings provided by the machine manufacturer. Moreover, based on this study and existing data from the literature, it can be inferred that without additional post-processing, or other supplementary procedures related to the process (such as the re-exposure of layers), achieving values significantly higher than approximately 120–130 HV is not feasible. Furthermore, when comparing the values of additively manufactured components with conventionally manufactured components from the same material, the AlSi7Mg0.6 aluminum alloy conventionally manufactured and achieved hardness levels above 120 HV only after additional heat treatment, precisely in the T6 state [35]. Hence, in the case of components manufactured using PBF-LB/M from aluminum alloys, it is feasible to eliminate certain processes associated with heat treatment if the demonstrated level of hardness is desirable.

### 3.4. Statistical Model

Based on the analysis of 22 parameter groups, analytical calculations were conducted to determine the regression coefficients $\beta_m$ and $\beta_{mn}$ (for m = 1,2,3; n = 1,2,3), along with the determination of the parameter value p. This facilitated the derivation of the final form of the mathematical model (quadratic surface regression), as represented by Equation (3). Furthermore, the coefficient of correlation $R^2$ was determined to be 0.85.

$$y = 6.9490 - 0.0735x_1 + 0.0078x_2 - 23.9161x_3 + 0.0003x_1^2 + 0.000003x_2^2 + 15.9355x_3^2 - 0.0001x_1x_2 + 0.0827x_1x_3 - 0.0088x_2x_3 + 0.4390 \quad (3)$$

In Table 5, the measured porosity values for the modeled elements produced using 37 parameter groups are presented. Additionally, the model, which indicated values for the 22 considered parameter groups in the DoE analysis, was also juxtaposed. The rest of the parameter groups (2, 4–8, 10–13, 26–28, 31–32) were taken for the validation of the created statistical model. The largest disparity between the mathematical model and the actual value occurred in group 35, with a difference of 0.36%. The mathematical model identified group 14 (out of 22 considered) as the parameter set, resulting in the lowest porosity for the modeled elements, aligning with actual measurements. An important consideration is that porosity measurement results for samples manufactured using the same energy density from different parameter groups exhibited variation. This clearly signals that when specifying manufacturing parameters for particular materials, it is crucial to provide not only energy density and layer thickness but also other parameters, such as exposure velocity, laser power, and the distance between hatching lines.

The mathematical model also enabled the generation of response surfaces (Figure 15), assuming that one value of the independent variable remained constant. The selected constant value was the distance between irradiation vectors, as this is the parameter least frequently subject to adjustments among the three considered. Below are presented five response surfaces for five values of the distance between irradiation vectors, which were among the 22 groups under consideration.

The remaining values of the two variables on the *X*-axis (laser power) and *Y*-axis (exposure velocity) were constrained within the predetermined upper limits. Specifically, these were 400 W due to the maximum achievable power on the machine used to produce the samples and 2000 mm/s, as higher values proved impractical in the process due to the phenomenon of 'balling', which may occur at such high speeds.

The orientations of fields indicating the lowest porosity values (<1%) exhibited a similar configuration in each case. They fell within the range of exposure velocities from 1200 mm/s to 2000 mm/s, while the laser power values ranged approximately from 225 W to 375 W. Each surface response distinctly delineated a technological window enabling the fabrication of elements without process failure risks. The position of this technological window, contingent upon the distance of values between irradiation vectors, is inclined at various angles to the *X*-axis.

This inclination increases with higher distance values between irradiation vectors. As the distance between laser paths increases, the power density decreases, consequently shifting the window model towards higher laser power values. It is crucial to note that the model's extrapolation of values beyond the considered parameter range in the 22 groups introduces significant error. Moreover, the theoretical minimum of function (4) represents a group of parameters, ostensibly allowing for the fabrication of elements with a porosity close to 0%. This group is detailed in Table 6.

**Table 5.** Measurements of porosity. Maximum and average perimeters and porosity predicted by the statistical model.

| Sample | Measured Porosity (%) | (Pρ) [%]—Statistic Model Equation (3) | Measured Max Perimeter [mm] | Measured Average Perimeter [mm] |
|---|---|---|---|---|
| 1 | 0.75 | 0.72 | 0.33 | 0.05 |
| 2 | 0.78 | - | 0.35 | 0.05 |
| 3 | 0.70 | 0.77 | 0.35 | 0.05 |
| 4 | 0.75 | - | 0.36 | 0.06 |
| 5 | 0.90 | - | 0.48 | 0.06 |
| 6 | 1.01 | - | 0.42 | 0.05 |
| 7 | 0.92 | - | 0.37 | 0.06 |
| 8 | 0.71 | - | 0.31 | 0.05 |
| 9 | 0.87 | 0.83 | 0.44 | 0.06 |
| 10 | 1.04 | - | 0.43 | 0.06 |
| 11 | 0.79 | - | 0.45 | 0.06 |
| 12 | 1.16 | - | 0.38 | 0.05 |
| 13 | 1.45 | - | 0.53 | 0.06 |
| 14 | 0.38 | 0.43 | 0.63 | 0.04 |
| 15 | 0.48 | 0.47 | 0.39 | 0.05 |
| 16 | 0.61 | 0.57 | 1.41 | 0.06 |
| 17 | 0.46 | 0.53 | 0.84 | 0.05 |
| 18 | 1.05 | 0.70 | 0.97 | 0.06 |
| 19 | 0.64 | 0.93 | 0.59 | 0.06 |
| 20 | 0.77 | 0.66 | 0.64 | 0.06 |
| 21 | 0.86 | 0.85 | 1.47 | 0.06 |
| 22 | 1.09 | 1.09 | 1.00 | 0.06 |
| 23 | 0.74 | 0.90 | 0.56 | 0.06 |
| 24 | 1.21 | 1.22 | 3.71 | 0.06 |
| 25 | 1.67 | 1.59 | 0.77 | 0.06 |
| 26 | 1.34 | - | 0.93 | 0.06 |
| 27 | 1.06 | - | 0.82 | 0.06 |
| 28 | 0.98 | - | 0.56 | 0.06 |
| 29 | 1.71 | 1.54 | 1.29 | 0.08 |
| 30 | 1.62 | 1.71 | 1.62 | 0.09 |
| 31 | 1.32 | - | 1.04 | 0.07 |
| 32 | 1.42 | - | 1.70 | 0.09 |
| 33 | 1.25 | 1.38 | 1.17 | 0.08 |
| 34 | 1.27 | 1.23 | 0.85 | 0.08 |
| 35 | 0.80 | 1.16 | 0.57 | 0.07 |
| 36 | 1.53 | 1.29 | 2.02 | 0.10 |
| 37 | 1.28 | 1.16 | 1.35 | 0.08 |

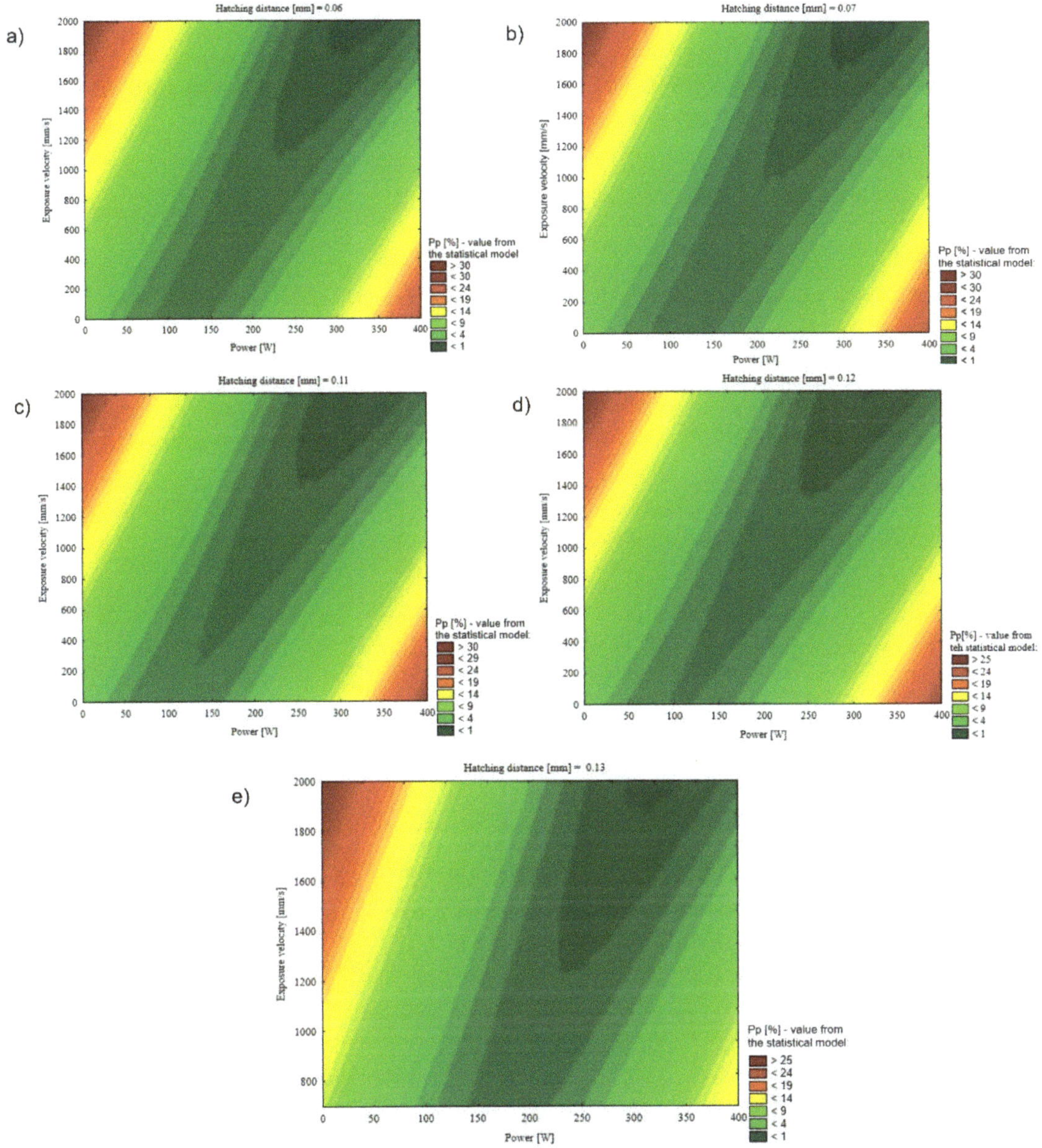


**Figure 15.** Surface response maps of the quadratic regression model for the constant hatching distance: (**a**) 0.06, (**b**) 0.07, (**c**) 0.11, (**d**) 0.12, and (**e**) 0.13.

**Table 6.** Predicted group of parameters using a function (4)—minimum of the function.

| Laser Power [W] | Exposure Velocity [mm/s] | Hatching Distance [mm] | Layer Height [mm] | Predicted Porosity [%] |
|---|---|---|---|---|
| 282.86 | 1650 | 0.13 | 0.03 | ≈0% |

The power density achieved from the proposed set of parameters amounted to 43.95 J/mm$^3$. Samples produced at similar power density values exhibited porosity levels of 0.79% and 1.16%. However, these were groups characterized by distinct values of individual components. Due to the limited scope of this article, this particular group will be further examined in subsequent research endeavors related to its utilization.

## 4. Conclusions

A properly adjusted process is the key to efficient and economical production and printing success, but above all, the quality of the microstructure and optimal mechanical properties of the printed parts. A wide range of process parameters of the PBF-LB/M technology for aluminum alloy AlSi7Mg0.6 were compared, and a mathematical model was developed to simplify the search for optimal parameters. Based on the obtained results, the following conclusions were drawn:

1. Using high laser power (>300 W), there were no unmelted particles of material; on the other hand, thermal cracks were found.
2. When using low laser power settings (<200 W) and high energy densities, paradoxically, a large number of large pores formed, and incorrectly melted powder particles also formed.
3. Microhardness was not affected by the change in process parameters and reached a very narrow range of values.
4. The quadratic surface regression model perfectly illustrates the technological windows and shows certain ranges of parameter values at which the porosity of the produced samples was low.
5. Within the considered range of 22 groups of parameters, regarding the group of manufacturing parameters for which the samples showed the lowest porosity, the model indicated the same group, No. 14, which was characterized by the smallest number of defects in the case of actual measurements.

**Author Contributions:** Conceptualization, J.K. and T.D.; methodology, J.K., J.Ł., T.D. and R.K.; software, Z.J. and K.J.; validation, J.K., T.D. and R.K.; formal analysis, T.D., R.K., Z.J. and K.J.; investigation, T.D., J.Ł. and J.K.; resources, J.K.; data curation, J.Ł., T.D. and K.J.; writing—original draft preparation, T.D., R.K. and J.Ł.; writing—review and editing, T.D. and J.K.; visualization, J.Ł., R.K. and T.D.; supervision, J.K. and Z.J.; project administration, J.K.; funding acquisition, J.K. All authors have read and agreed to the published version of the manuscript.

**Funding:** The research was funded by the Military University of Technology, grant number: 22-708.

**Institutional Review Board Statement:** Not applicable.

**Informed Consent Statement:** Not applicable.

**Data Availability Statement:** The research data are available upon request (via janusz.kluczynski@wat.edu.pl).

**Acknowledgments:** A part of this research is related to the specific research project 2024 "SV24-216" at the Department of Mechanical Engineering, University of Defense in Brno. This article was also based on research conducted during Tomáš Dražan's (from the University of Defense in Brno/Czech Republic) internship at the Military University of Technology in Warsaw/Poland. The internship was supervised by Janusz Kluczyński.

**Conflicts of Interest:** The authors declare no conflicts of interest.

## References

1. Martucci, A.; Aversa, A.; Bondioli, F.; Fino, P.; Lombardi, M. Synergic Strategies to Improve the PBF-LB\M Processability of Cracking-Sensitive Alloy. *Mater. Des.* **2022**, *224*, 111396. [CrossRef]
2. Kotadia, H.R.; Gibbons, G.; Das, A.; Howes, P.D. A Review of Laser Powder Bed Fusion Additive Manufacturing of Aluminiu Alloys: Microstructure and Properties. *Addit. Manuf.* **2021**, *46*, 102155. [CrossRef]

Smolina, I.; Gruber, K.; Pawlak, A.; Ziółkowski, G.; Grochowska, E.; Schob, D.; Kobiela, K.; Roszak, R.; Ziegenhorn, M.; Kurzynowski, T. Influence of the AlSi7Mg0.6 Aluminium Alloy Powder Reuse on the Quality and Mechanical Properties of LPBF Samples. *Materials* **2022**, *15*, 5019. [CrossRef] [PubMed]
Cacace, S.; Gökhan Demir, A.; Sala, G.; Mattia Grande, A. Influence of Production Batch Related Parameters on Static and Fatigue Resistance of LPBF Produced AlSi7Mg0.6. *Int. J. Fatigue* **2022**, *165*, 107227. [CrossRef]
Grande, A.M.; Cacace, S.; Demir, A.G.; Sala, G. Fracture and Fatigue Behaviour of AlSi7Mg0. 6 Produced by Selective Laser Melting: Effects of Thermal-Treatments. In Proceedings of the 25th Conference of the Italian Association of Aeronautics and Astronautics (AIDAA 2019), Rome, Italy, 9–12 September 2019; pp. 1138–1144.
Arvieu, C.; Galy, C.; Le Guen, E.; Lacoste, E. Relative Density of SLM-Produced Aluminum Alloy Parts: Interpretation of Results. *J. Manuf. Mater. Process.* **2020**, *4*, 83. [CrossRef]
Galy, C.; Le Guen, E.; Lacoste, E.; Arvieu, C. Main Defects Observed in Aluminum Alloy Parts Produced by SLM: From Causes to Consequences. *Addit. Manuf.* **2018**, *22*, 165–175. [CrossRef]
Fielden-Stewart, Z.; Coope, T.; Bacheva, D.; Kim, B.C. Effect of the Surface Morphology of SLM Printed Aluminium on the Interfacial Fracture Toughness of Metal-Composite Hybrid Joints. *Int. J. Adhes. Adhes.* **2021**, *105*, 102779. [CrossRef]
Aboulkhair, N.T.; Simonelli, M.; Parry, L.; Ashcroft, I.; Tuck, C.; Hague, R. 3D Printing of Aluminium Alloys: Additive Manufacturing of Aluminium Alloys Using Selective Laser Melting. *Prog. Mater. Sci.* **2019**, *106*, 100578. [CrossRef]
Cabrera-Correa, L.; González-Rovira, L.; de Dios López-Castro, J.; Botana, F.J. Pitting and Intergranular Corrosion of Scalmalloy® Aluminium Alloy Additively Manufactured by Selective Laser Melting (SLM). *Corros. Sci.* **2022**, *201*, 110273. [CrossRef]
Kumar, M.S.; Javidrad, H.R.; Shanmugam, R.; Ramoni, M.; Adediran, A.A.; Pruncu, C.I. Impact of Print Orientation on Morphological and Mechanical Properties of L-PBF Based AlSi7Mg Parts for Aerospace Applications. *Silicon* **2022**, *14*, 7083–7097. [CrossRef]
Fan, H.; Witvrouw, A.; Wolf-Monheim, F.; Souschek, R.; Yang, S. Hybrid Additive Manufacturing of Cost-Effective Automobile Parts by Die-Casting (AlSi7Mg0.3 (A356)) and Selective Laser Melting (AlSi7Mg0.5 (A357))-Effects of Substrate Surface Treatments on Bimetallic Interface Bonding Quality. Available online: https://osf.io/preprints/osf/exbsd (accessed on 16 June 2023).
Yadav, P.; Rigo, O.; Arvieu, C.; Lacoste, E. Microstructural and Mechanical Aspects of AlSi7Mg0.6 Alloy Related to Scanning Strategies in L-PBF. *Int. J. Adv. Manuf. Technol.* **2022**, *120*, 6205–6223. [CrossRef]
Mauduit, A.; Gransac, H.; Auguste, P.; Pillot, S.; Diószegi, A. Study of AlSi7Mg0.6 Alloy by Selective Laser Melting: Mechanical Properties, Microstructure, Heat Treatment. *J. Cast. Mater. Eng.* **2019**, *3*, 1–13. [CrossRef]
Godec, D.; Gonzalez-Gutierrez, J.; Nordin, A.; Pei, E.; Ureña Alcázar, J. *A Guide to Additive Manufacturing*; Springer Nature: Berlin/Heidelberg, Germany, 2022; ISBN 978-3-031-05863-9.
Wang, L.; Jiang, X.; Zhu, Y.; Ding, Z.; Zhu, X.; Sun, J.; Yan, B. Investigation of Performance and Residual Stress Generation of AlSi10Mg Processed by Selective Laser Melting. *Adv. Mater. Sci. Eng.* **2018**, *2018*, 7814039. [CrossRef]
Del Guercio, G.; Simonelli, M. Increasing the Build Rate of High-Strength Aluminium Alloys Produced by Laser Powder Bed Fusion. *Opt. Laser Technol.* **2023**, *161*, 109133. [CrossRef]
Bonneric, M.; Brugger, C.; Saintier, N. Effect of Hot Isostatic Pressing on the Critical Defect Size Distribution in AlSi7Mg0.6 Alloy Obtained by Selective Laser Melting. *Int. J. Fatigue* **2020**, *140*, 105797. [CrossRef]
Mauduit, A.; Gransac, H.; Auguste, P.; Pillot, S.; Diószegi, A.; Aboulkhair, N.T.; Simonelli, M.; Parry, L.; Ashcroft, I.; Tuck, C.; et al. Effect of hot isostatic pressing on the critical defect size distribution in AlSi7Mg0.6 alloy obtained by selective laser melting. *Int. J. Fatigue* **2022**, *15*, 6753–6760. [CrossRef]
Ma, Q.P.; Mesicek, J.; Fojtik, F.; Hajnys, J.; Krpec, P.; Pagac, M.; Petru, J. Residual Stress Build-Up in Aluminum Parts Fabricated with SLM Technology Using the Bridge Curvature Method. *Materials* **2022**, *15*, 6057. [CrossRef] [PubMed]
Dietrich, S.; Zaeh, M.F. Arc-Based Powder Production of AlSi7Mg0.6. *Procedia Manuf.* **2019**, *40*, 27–31. [CrossRef]
Ngo, T.D.; Kashani, A.; Imbalzano, G.; Nguyen, K.T.Q.; Hui, D. Additive Manufacturing (3D Printing): A Review of Materials, Methods, Applications and Challenges. *Compos. Part B Eng.* **2018**, *143*, 172–196. [CrossRef]
WWang, D.; Deng, G.W.; Yang, Y.Q.; Chen, J.; Wu, W.H.; Wang, H.L.; Tan, C.L. Interface Microstructure and Mechanical Properties of Selective Laser Melted Multilayer Functionally Graded Materials. *J. Cent. South Univ.* **2021**, *28*, 1155–1169. [CrossRef]
SLM Solutions AG Material Data Sheet Al-Alloy AlSi7Mg0,6/EN AC-42200. 2023, 7–9. Available online: https://www.slm-solutions.com/fileadmin/Content/Powder/MDS/MDS_Al-Alloy_AlSi7Mg0_6_0219_EN.pdf (accessed on 16 June 2023).
Adegoke, O.; Andersson, J.; Brodin, H.; Pederson, R. Influence of Laser Powder Bed Fusion Process Parameters on Voids, Cracks, and Microhardness of Nickel-Based Superalloy Alloy 247LC. *Materials* **2020**, *13*, 3770. [CrossRef] [PubMed]
Łuszczek, J.; Śnieżek, L.; Grzelak, K.; Kluczyński, J.; Torzewski, J.; Szachogłuchowicz, I.; Wachowski, M.; Karpiński, M. Processability of 21NiCrMo2 Steel Using the Laser Powder Bed Fusion: Selection of Process Parameters and Resulting Mechanical Properties. *Materials* **2022**, *15*, 8972. [CrossRef] [PubMed]
Chugunov, S.; Adams, N.A.; Akhatov, I. Evolution of SLA-Based Al2O3 Microstructure During Additive Manufacturing Process. *Materials* **2020**, *13*, 3928. [CrossRef]
Liu, X.; Zhang, M.; Li, S.; Si, L.; Peng, J.; Hu, Y. Mechanical Property Parametric Appraisal of Fused Deposition Modeling Parts Based on the Gray Taguchi Method. *Int. J. Adv. Manuf. Technol.* **2017**, *89*, 2387–2397. [CrossRef]
Gaulard, F.; Schmied, J.; Fuchs, A.; Major, B.; Sandu, A.V.; Abdullah, M.M.A.B.; Nabiałek, M.; Tański, T.; Zieliński, A.; Fiuk, G.; et al. Sustainability in Production in the Aspect. *Bull. Polish Acad. Sci. Technol. Sci.* **2021**, *69*, 1–9. [CrossRef]

30. Pinkerton, A.J. [INVITED] Lasers in Additive Manufacturing. *Opt. Laser Technol.* **2016**, *78*, 25–32. [CrossRef]
31. Kair, A.B.; Sofos, K. Additive Manufacturing and Production of Metallic Parts in Automotive Industry: A Case Study on Technic Economic and Environmental Sustainability Aspects. 2014. Available online: http://www.diva-portal.org/smash/record.js pid=diva2:740682&dswid=-4901 (accessed on 26 June 2023).
32. Andreau, O.; Pessard, E.; Koutiri, I.; Penot, J.D.; Dupuy, C.; Saintier, N.; Peyre, P. A Competition between the Contour ar Hatching Zones on the High Cycle Fatigue Behaviour of a 316L Stainless Steel: Analyzed Using X-Ray Computed Tomograp *Mater. Sci. Eng. A* **2019**, *757*, 146–159. [CrossRef]
33. Masuo, H.; Tanaka, Y.; Morokoshi, S.; Yagura, H.; Uchida, T.; Yamamoto, Y.; Murakami, Y. Effects of Defects, Surface Roughne and HIP on Fatigue Strength of Ti-6Al-4V Manufactured by Additive Manufacturing. *Procedia Struct. Integr.* **2017**, *7*, 19– [CrossRef]
34. Vaudreuil, S.; Bencaid, S.E.; Vanaei, H.R.; El Magri, A. Effects of Power and Laser Speed on the Mechanical Properties AlSi7Mg0.6 Manufactured by Laser Powder Bed Fusion. *Materials* **2022**, *15*, 8640. [CrossRef]
35. Birol, Y. Effect of Solution Heat Treatment on the Age Hardening Capacity of Dendritic and Globular AlSi7Mg0.6 Alloys. *Int. Mater. Res.* **2010**, *101*, 439–444. [CrossRef]

materials

icle

# ffect of Hardness Distribution on Strength of Narrow-Gap :ot-Wire Laser-Welded Joint for High-Tensile Strength Steel

kkapun Greebmalai, Kazuyuki Matsumoto, Keita Marumoto and Motomichi Yamamoto *

Graduate School of Advanced Science and Engineering, Hiroshima University, Hiroshima 739-8527, Japan; d225295@hiroshima-u.ac.jp (J.G.)
* Correspondence: motoyama@hiroshima-u.ac.jp

**Abstract:** Application of high-heat input welding on high-tensile strength steels causes deterioration of mechanical properties of the welded joint, due to softening and grain coarsening in the heat-affected zone (HAZ). In this study, low-heat input narrow-gap hot-wire laser welding was applied to 12 mm thick 780 MPa-class high-tensile strength steel plate. Conditions were optimized based on microstructural observations of joints produced at various welding speeds. Heat input was estimated from measured grain size. Evaluation of properties of joints welded at 0.5 m/min revealed sound toughness, tensile strength, and elongation. The effect of undermatched weld metal width on joint strength was analyzed using a finite element method. When the width of undermatched weld metal was 2.5 mm, the joint strength was 99% of the base metal strength; when it was 7.5 mm, the strength dropped to 95%. The effect of HAZ softening width on joint strength with even-matched weld metals was similarly analyzed, showing that even when the HAZ softening width was 2.0 mm, the joint strength was 98% of the base metal strength. The results of this study suggest that narrow-gap hot-wire laser welding can efficiently reduce heat input and the HAZ softening zone, thereby achieving both high strength and high toughness.

**Keywords:** laser welding; hot wire; narrow gap; high-tensile strength steel; HAZ softening

## 1. Introduction

High-efficiency welding can be achieved by minimizing the number of passes, although a lower number of welded passes requires a higher material deposition ratio and higher heat input. During high-heat input welding, the heat-affected zone (HAZ) of the welded joint experiences grain coarsening proportional to the thermal effects of the higher heat input, which leads to deterioration in the mechanical properties of the joint. Studies have reported negative impacts on high-tensile strength steels during the welding process due to this heat-induced behavior [1–4]. Reducing the heat input is considered an effective method for mitigating joint deterioration; however, a lower heat input may lead to a decrease in welding efficiency.

Narrow-gap welding is a method characterized by using a small groove width with a small bevel angle, resulting in reductions in the amounts of deposition material and heat input, thereby enabling high-efficiency welding [5,6]. In this process, gas metal arc welding (GMAW) and gas tungsten arc welding (GTAW) require the torch to access the interior of the groove near the welding area. This access limitation means that the groove width cannot be too narrow. In contrast, laser welding can effectively reach and weld within the confines of an extremely narrow groove [7–11], benefiting from its small beam profile. In terms of material deposition or wire insertion, the hot-wire method involves feeding an electrically heated wire into the molten pool during the welding process. With the external heat source heating the filler wire, less additional heat from the arc or laser is needed to completely melt it into the molten pool. This technique allows for a higher amount of filler deposition, facilitating high-efficiency welding. Moreover, recent years have seen ongoing studies on hot-wire narrow-gap laser welding aimed at enhancing the efficiency of narrow-gap welding while maintaining lower heat input [12–16].

This study combined narrow-gap laser welding and the hot-wire method, aimed at achieving low-heat input welding, which reduces the heat-induced microstructural impact on the HAZ. In addition, an attempt was made to control the strength (hardness) and toughness of the weld metal by using a softer wire compared to the strength of the base steel, making use of the large degree of freedom in wire selection of the hot-wire method and the large cooling rate of this welding method.

Low-heat input welding has a higher cooling rate than high-heat input conditions [17,18] and hardness of low-heat input weld metal increases [18], which could lead to increased cracking risk [19,20]. To prevent crack initiation, a stress-reduction method involving the selection of a low-strength filler wire was implemented in this study of hot-wire narrow-gap laser welding, particularly for the welding of thick high-strength plates [21,22]. This approach results in a lower strength of the weld metal than that of the base material, leading to undermatched strength and a consequent decrease in joint strength. Both undermatched weldment strength and HAZ softening can diminish the overall strength of the joint.

Hochhauser et al. [23] conducted strength evaluations on thermo-mechanically controlled process (TMCP) high-tensile steels with varying widths of the softening zone, revealing no difference in base-metal strength when the softening zone width was 0.25 times the plate thickness. Similarly, Rodrigues et al. [24] reported that there was no difference in base-metal strength when the softening zone width was less than one-third of the plate thickness in high-tensile steels. Therefore, with the application of plastic constraints, it is possible to achieve high joint efficiency, even when the joint has a softening region. When the narrow-gap concept is carried out using the hot-wire laser welding method, which leads to low heat input and a narrow width of the weld metal and HAZ, it will be possible to achieve a welded joint with sufficient strength or equivalent to that of the base material despite undermatched strength of the filler wire.

This study clarifies the effects of heat input conditions on the microstructure, occurrence of incomplete fusion, and hardness distribution of joints welded using this technique. Additionally, evaluations of impact absorption energy and tensile strength were conducted using joints welded under optimized conditions. Welded joints with varying weld metal and softening widths were assessed using finite element method (FEM) analysis. Results regarding joint efficiency relative to weld metal and softening width are presented.

## 2. Materials and Methods

### 2.1. Materials and Experimental Method

In this study, 780 MPa-class high-tensile strength steels with plate thicknesses of 10 mm and 12 mm were used as the base metal. A narrow U-shaped groove was employed, with a 0.5 mm groove face and 1 mm radius at the bottom. The groove shape is shown in Figure 1. A 490 MPa-grade wire with 1.2 mm diameter was used as the filler material. Chemical compositions of the materials are shown in Table 1.

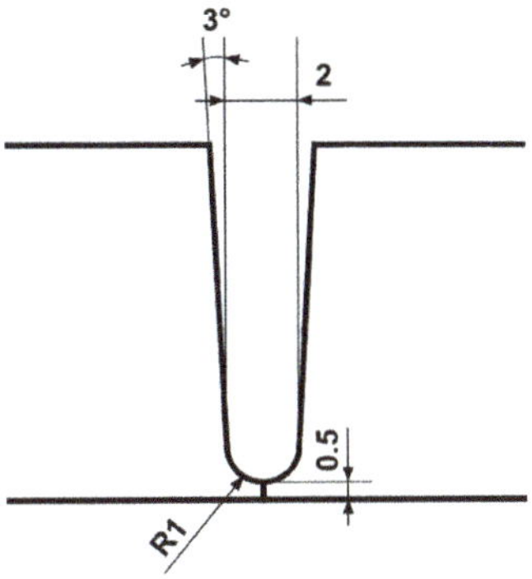


**Figure 1.** U-shaped narrow groove geometry (unit: mm).

**Table 1.** Chemical compositions of filler wire and base metal.

| Material | Chemical Composition, Mass% | | | | | | | | | |
|---|---|---|---|---|---|---|---|---|---|---|
| | C | Si | Mn | P | S | Cu | Cr | Nb | V | Ti |
| Base metal 780 MPa-class steel | 0.14 | 0.25 | 1.32 | 0.011 | 0.004 | 0.01 | 0.25 | 0.002 | 0.002 | 0.01 |
| Filler wire S Z3316W49A3U16 | 0.06 | 0.73 | 1.74 | 0.021 | 0.007 | 0.24 | 0.03 | -- | -- | -- |

The experimental setup is shown in Figure 2. Welding was performed with the laser beam leading, and the filler wire and shielding nozzle were installed behind the molten pool. A rectangular laser beam of 1.6 mm × 11.0 mm with a focusing and homogenizer lens configuration was used. The filler wire tip was located at the rear of the laser beam spot at 10 mm from the front of the beam profile. Argon was used as shielding gas following the hot-wire torch.

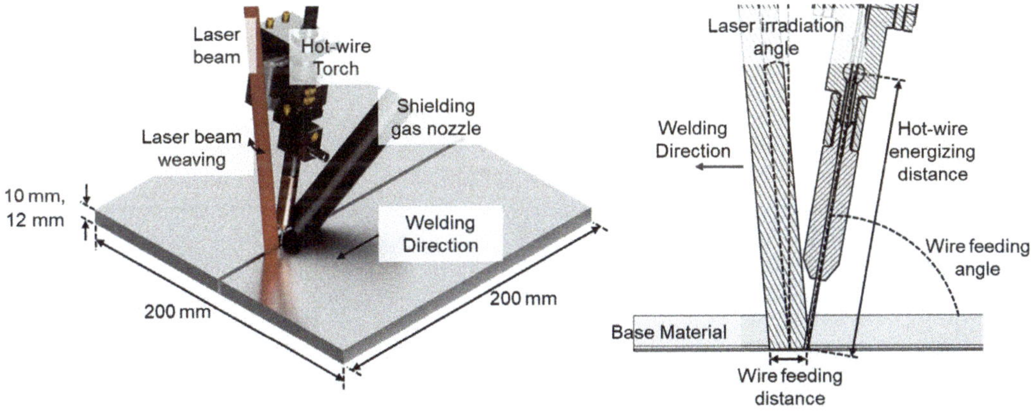


**Figure 2.** Experimental setup.

The length of the laser spot was oriented parallel to the welding direction. Energy density distribution of the laser beam is shown in Figure 3. In addition, a small laser beam with oscillatory motion was used for reaching from side to side of the narrow joint. The smoothing-square (s-square) oscillation waveform is shown in Figure 4. The time duration at high and low amplitudes (beam located at groove wall sides) was longer than that of a sine waveform. This characteristic assisted in suppression of the incomplete fusion that usually occurs at the groove walls of a narrow joint.

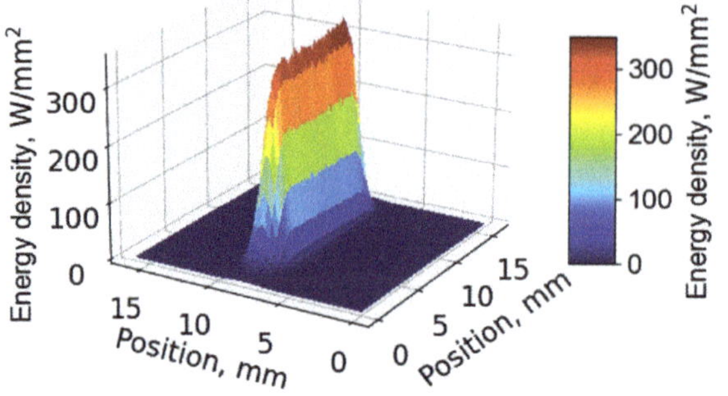


**Figure 3.** Energy distribution of laser beam spot.

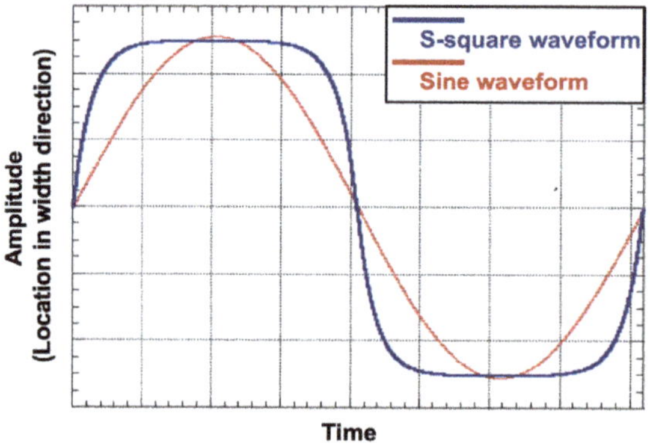


**Figure 4.** Weaving motion of laser beam spot.

Experiments were conducted under the conditions shown in Table 2 to investigate the effect of heat input by changing the welding speed from 0.3 m/min to 0.75 m/min on the occurrence of defects, hardness distribution, and grain size of the HAZ. The wire feeding speed was varied following welding speed to maintain the deposition rate per unit length. A diode laser oscillator was used as the main heat source, and the laser power was fixed at 6 kW. A 10 mm thick groove could be welded in one pass under these conditions. The hot-wire current was set to a suitable limiting current value for each wire feeding speed at which no fusing or fluctuation occurred during welding under the 80 mm energization distance for joule heating.

**Table 2.** Experimental conditions.

| Parameter | (a) | (b) | (c) |
|---|---|---|---|
| Laser power, kW | | 6.0 | |
| Laser spot, mm | | 1.6 × 11.0 | |
| Laser irradiation angle, deg | | 5 | |
| Defocus length, mm | | 10 | |
| Welding speed, m/min | 0.3 | 0.5 | 0.75 |
| Energization distance, mm | | 80 | |
| Wire feeding speed, m/min | 8.2 | 13.7 | 20.5 |
| Hot-wire current, A | 158 | 178 | 248 |
| Weaving frequency, Hz | | 10 | |

The welded joints were cut, and the length of the lack-of-fusion region was measured from three cross sections and represented as the lack-of-fusion ratio relative to the total groove length. Vickers hardness was measured in the middle of the joint thickness on the bead width direction.

The grain size was measured using an optical microscope and linear-intercept method. The measurement locations and methods are shown in Figure 5. The fine-grained region was not measured due to the difficulty of counting intercepts for the high-density grains.

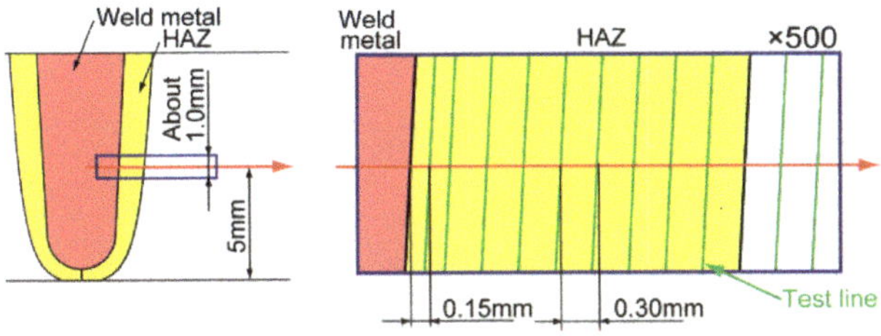


**Figure 5.** Welding and evaluation region.

The heat input of this welding method was estimated using the measured grain size by performing an inverse estimation using the grain growth equation from Ikawa et al. [25]. The grain growth equation for high-tensile steel is expressed as

$$\log_{10}\left(D^4 - D_0^4\right) = -92.64 + 2\log_{10}\eta' q' + \frac{1.291\times10^{-1}}{\left(\frac{l'}{\eta' q'}\right)+1.587\times10^{-3}}, \quad (1)$$

where $D$ is the grain diameter after welding, $D_0$ is the grain diameter of the base metal (27.3 μm), $\eta' q'$ is the amount of heat input, and $l'$ is the distance from the weld.

On the basis of the results of the varied welding speed experiments, the welding conditions were optimized for a test specimen with a plate thickness of 12 mm, and a Charpy impact test and tensile test were conducted. The welding conditions are shown in Table 3. The Charpy impact test was conducted using a V-notch test specimen at temperatures of −40 °C and −20 °C. In addition, experiments were conducted at two positions for the V-notch: the weld metal and fusion boundary. The dimensions of the test specimens and test positions are shown in Figure 6.

**Table 3.** Welding conditions for Charpy impact test and tensile test.

| Parameter | |
|---|---|
| Laser power, kW | 6.0 |
| Laser spot, mm | 1.6 × 11.0 |
| Laser irradiation angle, deg | 5 |
| Defocus length, mm | 0 |
| Welding speed, m/min | 0.5 |
| Energization distance, mm | 80 |
| Wire feeding speed, m/min | 12.7 |
| Hot-wire current, A | 180 |
| Weaving frequency, Hz | 15 |

The tensile test was conducted in accordance with the JIS standard [26], and the dimensions used were those of specimen No. 5, as specified in the JIS standard [26]. The shape of the tensile specimen is shown in Figure 7.

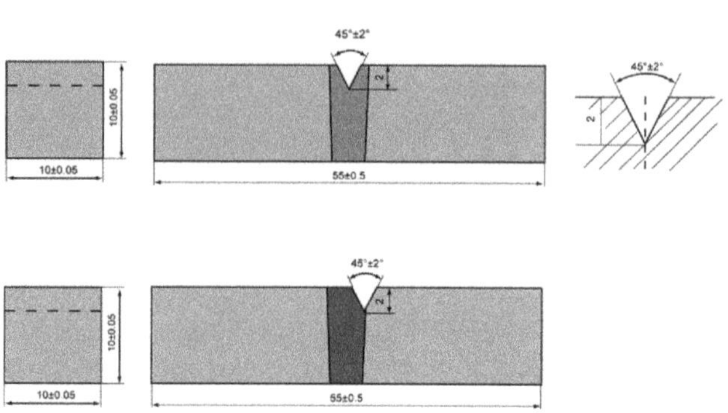


**Figure 6.** Charpy impact testing samples (unit: mm).

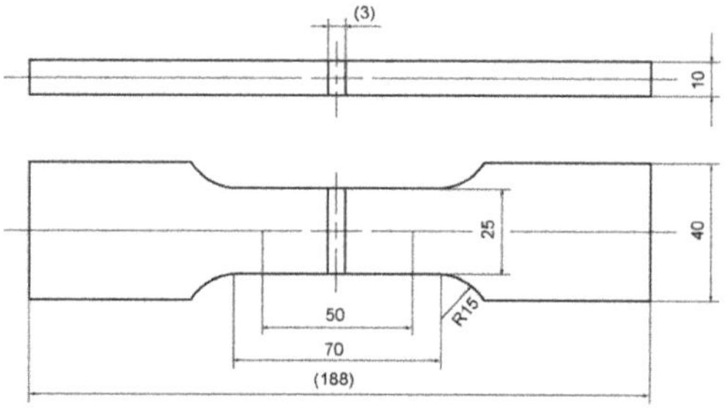


**Figure 7.** Tensile test specimen (unit: mm).

## 2.2. Finite Element Analysis Model

To clarify the effect of the width of the undermatched zone of the weld metal and softening HAZ zone, FEM analyses were conducted using tensile test models. The analysis was conducted on a one-quarter model using MARC. The region included the weld metal, softened zone of the HAZ, and base metal. The analysis model is shown in Figure 8. The boundary conditions were set to be symmetrical on the symmetry plane, and a forced variation of 0.0415 mm per step was applied in the tensile direction. The analysis was stopped when the maximum contact variation in the X-axis (force direction) reached 3 mm. The boundary conditions are shown in Figure 9.

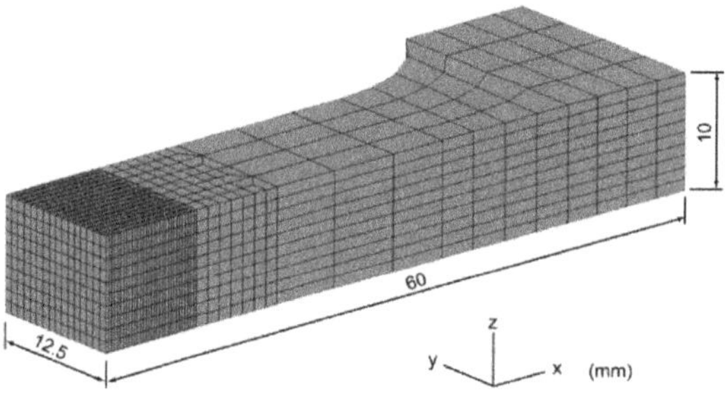


**Figure 8.** Finite element model grid.

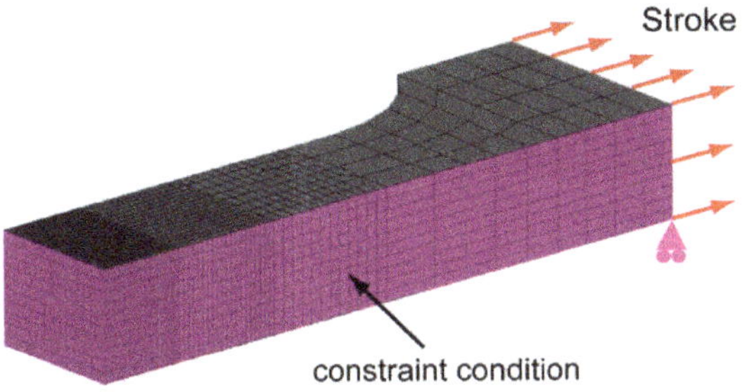


**Figure 9.** Boundary conditions and constraints of finite element model.

On the basis of hardness measurements of actual welded joints, the changes in mechanical properties were estimated to simulate a joint where the hardness changed between the weld metal, HAZ softening area, and base metal. Specifically, the mechanical properties were simulated by performing linear interpolation at each hardness level [27] based on the true stress–true strain relationships of 780 MPa-class high-tensile strength steel (base metal hardness: 280 HV) which is same as the used base material and 490 MPa-class high-tensile strength steel (base metal hardness: 170 HV) which is same as the used filler metal. The true stress–true strain relationships used for interpolation are shown in Figure 10. The hardened and softened zones were divided into regions of 0.5 mm each. An example of the hardness distribution set in the analysis model is shown in Figure 11.

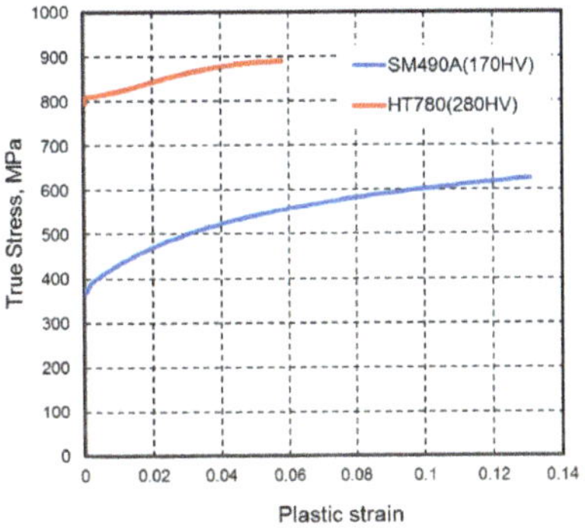


**Figure 10.** True strain–stress relation of 780 MPa-class and 490 MPa-class high-tensile strength steels.

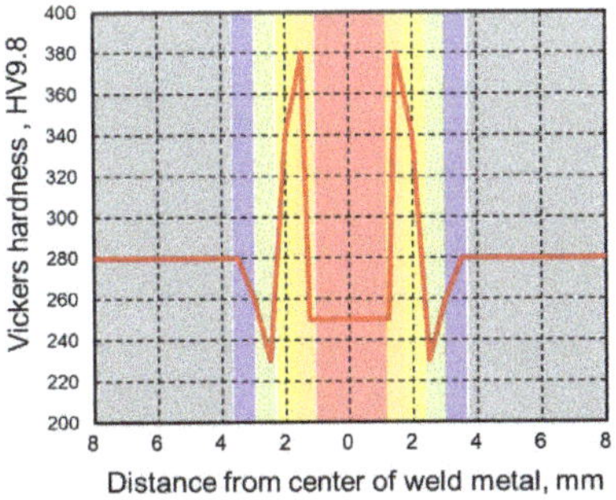


**Figure 11.** Example of hardness distribution of finite element model (red: weld metal, dark yellow~light yellow: hardened HAZ, green~blue: softened HAZ, gray: base metal).

The effect of weld metal width on the joint strength was analyzed by changing the weld metal width, which was considered when using low-strength wire. The hardness of the weld metal was set at 220 HV, which is softer than the base metal hardness of 280 HV. The width of the weld metal was set at three conditions of 2.5, 5.0, and 7.5 mm. The analysis conditions near the weld metal are shown in Table 4.

**Table 4.** Finite element analysis conditions for undermatched weld metal width effect.

| Region | Vickers Hardness, HV | Width of Each Region (Half Width), mm | | |
|---|---|---|---|---|
| Weld metal | 220 | 2.5 (1.25) | 5.0 (2.5) | 7.5 (3.75) |
| HAZ 1 | 380 | | 0.5 | |
| HAZ 2 | 340 | | 0.5 | |
| HAZ 3 | 230 | | 0.5 | |
| HAZ 4 | 260 | | 0.5 | |
| Base metal | 280 | | -- | |

The softened HAZ width was varied in an even-matched joint where the hardness of the weld metal was equivalent to the base metal hardness of 280 HV. The analysis was performed under three conditions with softened HAZ widths in region 3 of 1.0, 1.5, and 2.0 mm. The analysis conditions are shown in Table 5.

**Table 5.** Finite element model conditions for softened heat-affected zone width effect.

| Region | Vickers Hardness, HV | Width of Each Region (Half Width), mm | | |
|---|---|---|---|---|
| Weld metal | 280 | | 2.5 (1.25) | |
| HAZ 1 | 380 | | 0.5 | |
| HAZ 2 | 340 | | 0.5 | |
| HAZ 3 | 230 | 0.5 | 0.75 | 1.0 |
| HAZ 4 | 260 | | 0.5 | |
| Base metal | 280 | | -- | |

## 3. Results and Discussion

### *3.1. Effect of Welding Speed on Microstructure and Hardness Distribution*

The experimental conditions are detailed in Table 2. Cross-sectional photographs of the welded joints are presented in Figure 12. Qualitative observation indicates that the welded area and mid-bottom bead width across all conditions showed insignificant variation with changes in travel speed. The mid-bottom bead width was influenced by the welding speed, which aligns with findings from previous studies [28,29]. The top width of the welded joint showed slight reductions at travel speeds of 0.5 m/min compared with 0.3 m/min, while a more significant reduction was observed at 0.75 m/min due to the lower heat input under the higher welding speed condition [30].

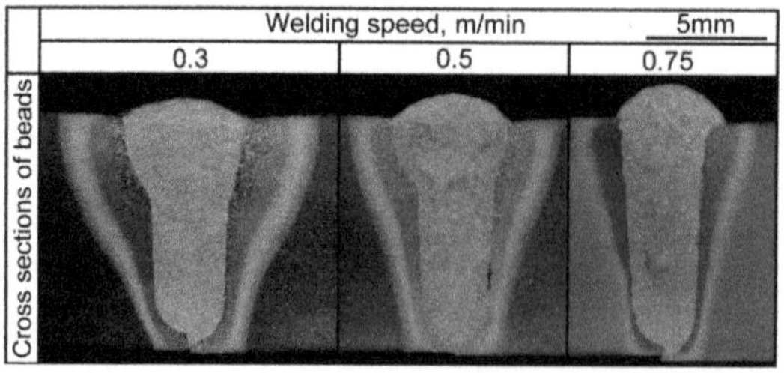


**Figure 12.** Cross sections under three welding speed conditions.

As shown in the cross section in Figure 12, a lack of fusion at the groove wall could be observed for travel speed conditions of 0.5 m/min and 0.75 m/min. This lack of fusion occurred due to insufficient heat input to balance the deposition rate, laser energy density, and welding speed [30,31]. The lack-of-fusion ratio at each welding speed was measured from the cross-sectional observations and is shown in Figure 13. No lack of fusion was observed at a welding speed of 0.3 m/min.

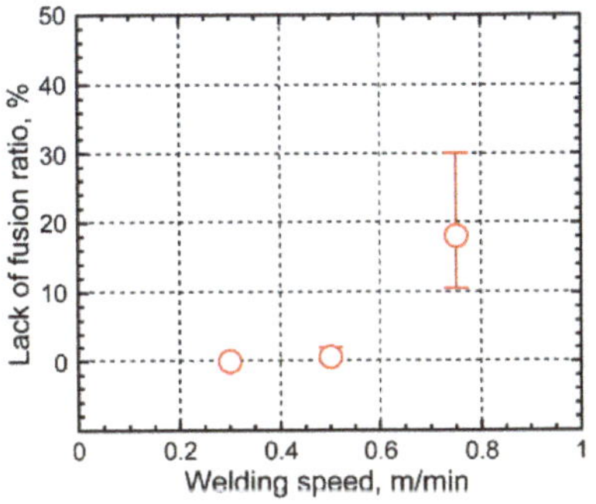


**Figure 13.** Lack-of-fusion ratios as a function of welding speed.

The microstructure was observed at the center of the plate thickness (5 mm from the bottom of the joint) on cross-sectional samples. Images obtained using an optical microscope are presented in Figure 14. The macroscopic images indicate that the HAZ width decreased with increasing welding speed, as confirmed in both Figures 12 and 14. The microstructure of the base metal, shown in location (h) for all conditions, exhibited a morphology characterized by gray polygonal grains with dark black stripes, which several studies have identified as tempered martensite [22,32–34]. Additionally, the presence of thin white elongated grains indicated the presence of bainite and ferrite [33,34]. In the welded metal at location (a) with a travel speed of 0.3 m/min, the microstructure revealed acicular ferrite and grain boundary ferrite (GBF) [33]. However, as the travel speed increased to 0.5 and 0.75 m/min, the microstructure showed a higher fraction of bainite and acicular ferrite, with a lower fraction of GBF [34]. Similar morphologies were observed at all travel speed conditions across the fusion line, HAZ, and base metal, although the width of this area decreased due to the reduction of heat input with increasing travel speed.

A large fraction of bainite was observed close to the fusion line at location (b), extending from location (c) to location (e), where the size of the bainite decreased. This behavior has been attributed to grain growth resulting from heat exposure, as reported in previous studies [34]. Small equiaxed grains were found at location (f), suggesting recrystallization of the metal [34]. Additionally, nucleation was observed at the grain boundaries at location (g), with high carbon diffusion occurring at these boundaries, for which similar behavior has been documented [33]. These results indicate that the material subjected to heat during the welding process consisted of recrystallized regions and areas of grain growth, suggesting that this region may have contributed to softening in the HAZ. Furthermore, the literature reports similar microstructural behavior and morphology, revealing both hardening and softening zones in the HAZ of welded joints. This variation in microstructure results in higher mechanical properties in these regions compared with those of the base metal [35,36]. However, further investigations are needed to fully understand the characteristics of high-strength steel joints.

The results of Vickers hardness measurements at the midpoint of the welded joint are shown in Figure 15. The hardness distribution exhibited a similar profile along the welded joint across all conditions, with increased hardness influenced by the refined grain structure during recrystallization at location (f) in Figure 14. The reduction in hardness within the HAZ was confirmed by the microstructure observed at location (g) in Figure 14.

This hardness distribution result aligns with findings from several studies on high-strength steel welding [22,32].

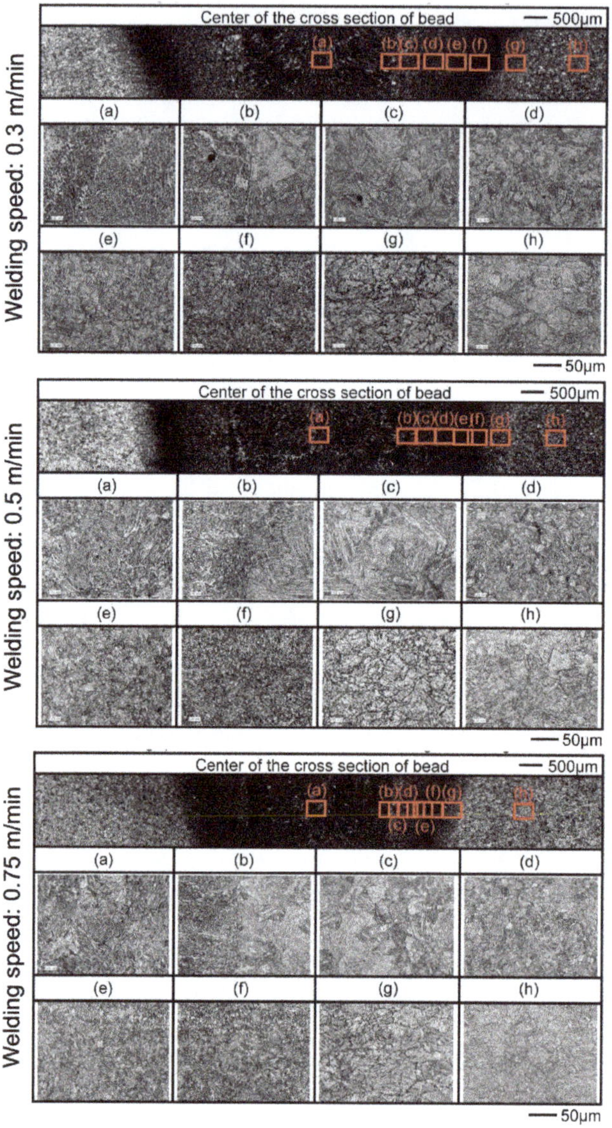


**Figure 14.** Microstructures of joints welded at various welding speed conditions.

Welding speeds of 0.3, 0.5, and 0.75 m/min resulted in hardness values of the welded metal of about 260, 300, and 300 HV, respectively. The hardness increase resulted from suppression of GBF at the 0.5 and 0.75 m/min travel speed conditions. Furthermore, the average hardness of the welded metal at the welding speed of 0.3 m/min was undermatched to base-metal hardness in this study.

The grain size measurements are presented in Figure 16. The grain size decreased from the fusion line to the base metal in all conditions [37,38]. As the travel speed increased, the grain size reduced due to the influence of lower heat input and shortening of the HAZ region [38]. Moreover, the region near the fusion boundary revealed large grain sizes of 100, 60, and 40 μm for travel speeds of 0.3, 0.45, and 0.75 m/min, respectively. In contrast, grain sizes in the base metal were smaller, indicating that this region can be defined as a coarsening area resulting from heat exposure during welding, which promotes grain growth [37,38].

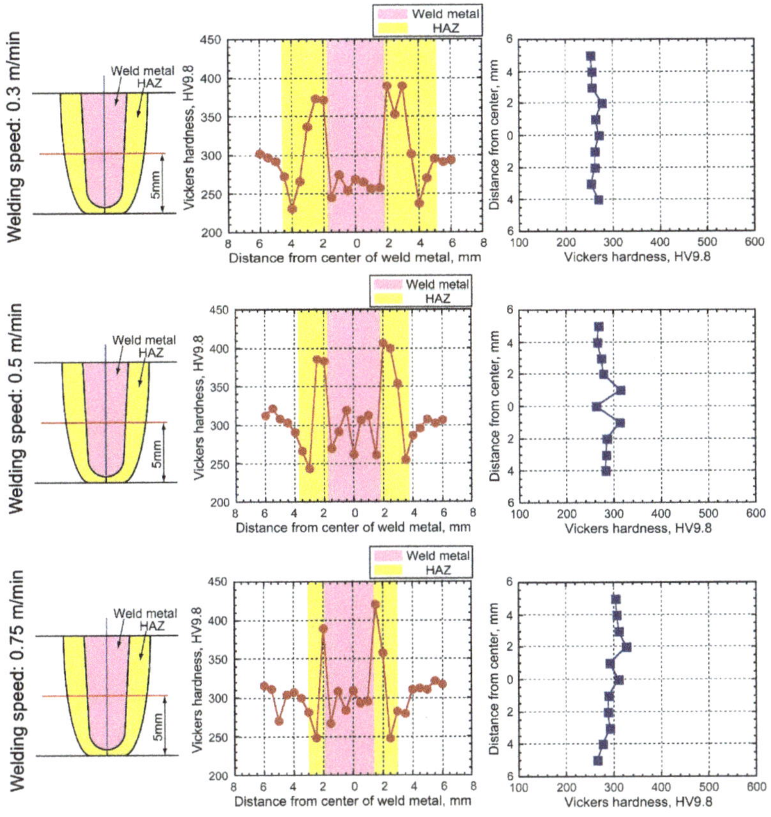


**Figure 15.** Vickers hardness distribution results.

The width of the grain-coarsening region was found to be proportional to the heat input. On the basis of these observations, the heat input was estimated using Equation (1). The heat input was calculated using the measured grain size results from Figure 16, taken at a distance of 0.15 mm from the fusion boundary [25], across all travel speed conditions, and substituted into Equation (1). The results of these calculations are shown in Figure 17. Comparison of the calculated values shows that the heat inputs were approximately 11.0 kJ/cm, 7.5 kJ/cm, and 4.5 kJ/cm at welding speeds of 0.3 m/min, 0.5 m/min, and 0.75 m/min, respectively, as indicated by the best fit to the experimental grain size results.

This hot-wire laser narrow-gap welding process is therefore considered a low-heat input welding method compared with conventional processes such as GMAW [39–42] or laser welding [43,44].

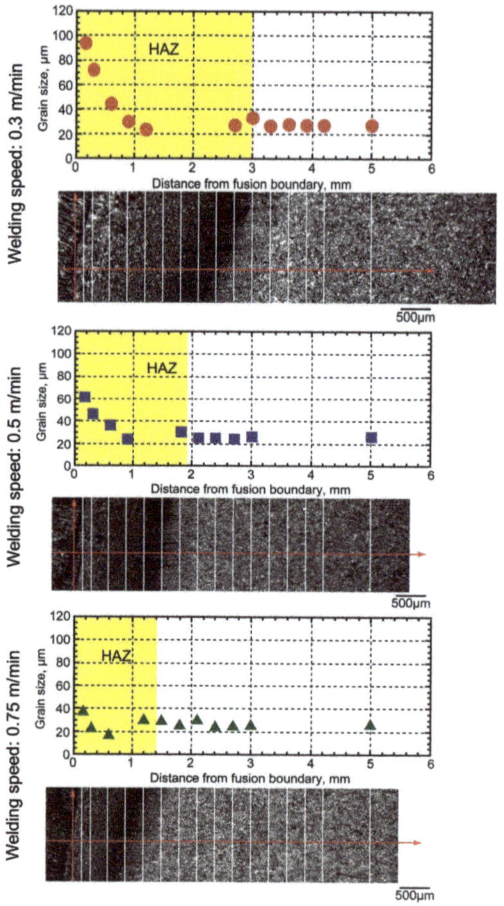


**Figure 16.** Measured grain size from fusion boundary.

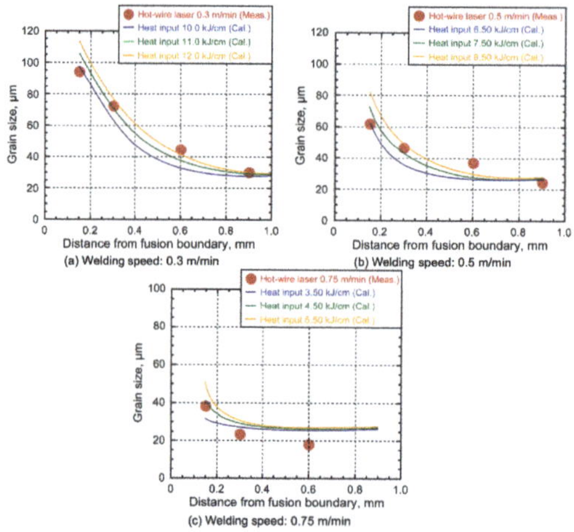


**Figure 17.** Comparison of calculation heat input and measured grain sizes.

### 3.2. Joint Evaluation of Mechanical Properties and Finite Element Model Analysis

Experiments were conducted at various welding speeds on material of 12 mm thickness to evaluate the mechanical properties. The conditions are listed in Table 3. Cross-sectional observation of bead formation of a butt joint is shown in Figure 18. The joint welded at a speed of 0.5 m/min was selected as the optimized condition due to being the highest speed at which uniform penetration with defect-free formation was achieved. Therefore, the condition of 0.5 m/min was selected for testing of mechanical properties, including Chappy impact and tensile tests.

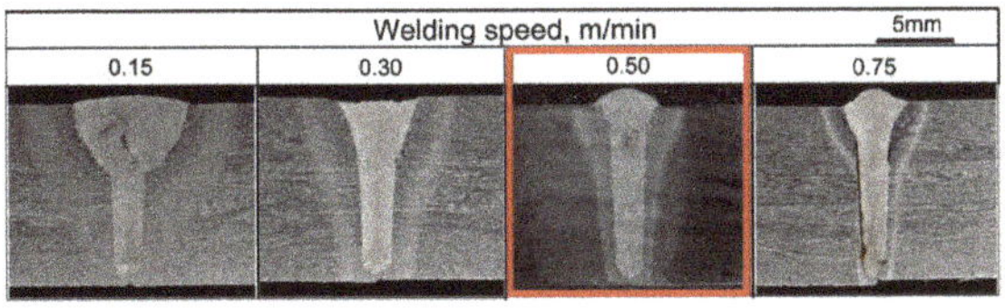


**Figure 18.** Cross-section appearances under various welding speed conditions (red: selected optimum welding speed).

The Charpy impact test was carried out at temperatures of −20 °C and −40 °C at the weld metal center and fusion boundary. The results are represented in Figure 19. At the positions of weld metal (a) and fusion boundary (b) during −20 °C testing, the result of 47 J satisfies the requirement for absorption energy higher than that of the joint, which is influenced by the microstructure [22,45] of a joint similar to that shown in Figure 14. Remark occurred at the fusion boundary due to higher absorption energy compared with that of the weld metal due to formation of a large tempered structure from heat subjection during welding [32]. Moreover, this area was exposed to large fracture path deviation (FPD) of the testing sample, evaluated close to the boundary of the weld metal and base metal areas. During testing at −40 °C, the joint energy absorption was shown to be below the designed value of 47 J for both the weld metal (a) and fusion boundary (b) areas. These were influenced by the microstructure comprising the larger bainite fraction in weld metal and temper microstructure near fusion boundary during this low-temperature testing [45]. However, selection of a filler wire with higher toughness or a post-heat treatment process might be considered to modify the microstructure in this weld metal area for better energy absorption [45].

Tensile testing was conducted on samples prepared from the base material and welded joint. The yield stress, ultimate tensile strength, and elongation were obtained. The welded joint was found to be fractured in the base metal area, while the overall welded joint strength was not affected by the hot-wire laser welding process. The appearance of the welded joint tensile sample after the testing is shown in Figure 20.

Results from evaluations of several metallurgy and mechanical properties suggested and confirmed that the welded joint exhibited a softening region in the HAZ, with the welded metal being softer than the base metal [22,33]. However, during tensile testing, failure of the welded joint was observed to occur in the base metal. This behavior, as noted in [46], is influenced by the narrow undermatched region of the welded joint (either in the welded metal or HAZ), which does not undervalue the overall deterioration of joint properties (tensile strength), resulting in joint fractures occurring in the base material region.

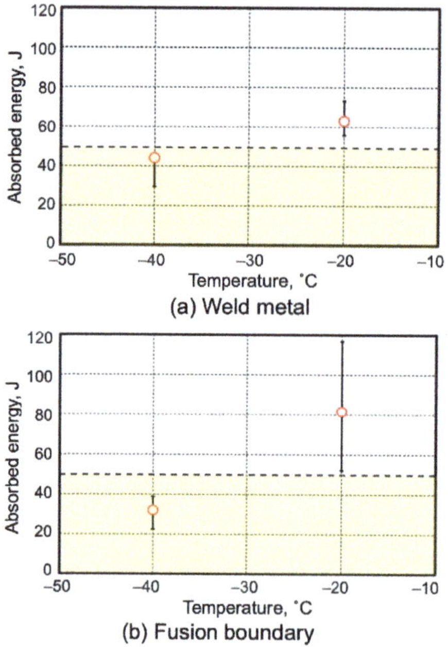


**Figure 19.** Charpy impact test results.

| | Base Material | Welded joint |
|---|---|---|
| Yield stress, MPa | 796 | 799 |
| Ultimate tensile strength, MPa | 841 | 851 |
| Elongation, % | 24.0 | 24.6 |

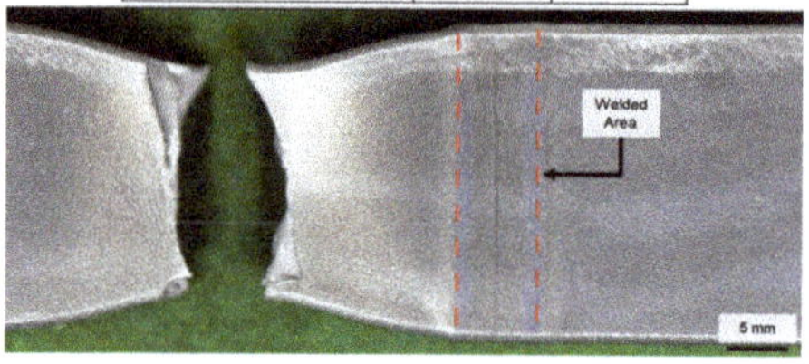


**Figure 20.** Tensile test result on welded joint (red dashed lines: fusion boundaries).

### 3.3. Finite Element Model Analysis Results

The effect of the weld metal width on the joint strength was analyzed for various weld metal widths using FEM models in which the welded joint was considered when low-strength wire was used (i.e., undermatched joints). Contour diagrams of the mean normal stress just after yielding (start of plastic deformation), during progression of plastic deformation (displacement at 0.415 mm), and at the end of the analysis (displacement at 3.00 mm) for each condition are shown in Table 4 and Figure 21. At the start of plastic deformation, the stress in the narrow range of the HAZ hardening zone was low, while that in the surrounding weld metal and HAZ softening zone was very high. As deformation progressed (displacement at 0.415 mm), this trend became more pronounced. At the end of the analysis, when high stress conditions prevail, displacement and necking were observed on the weld metal located at center of the specimen (left edge of the analysis model).

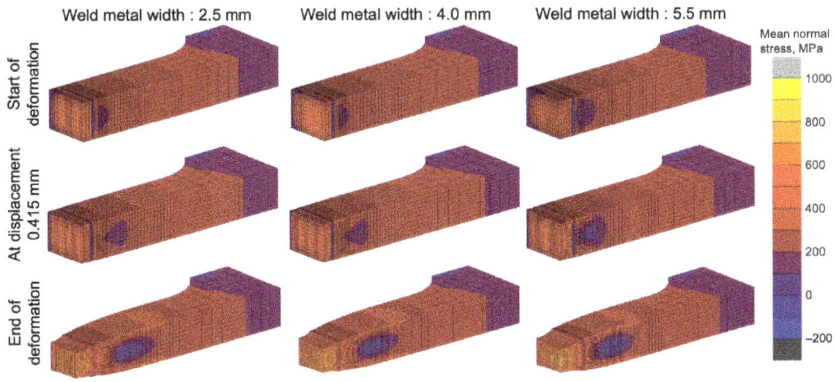


**Figure 21.** Mean nominal stress contours and deformation in analysis of undermatched weld metal width effect.

The joint nominal tensile strength ($\sigma_J$) obtained from FEM analysis was compared by normalization using the base material tensile strength ($\sigma_B$). The result is represented as joint efficiency in Figure 22. Although the welded metal width increased, the efficiency of the welded joint decreased, as reported by several other researchers who used FEM analyses [47,48]. However, the highest joint efficiency could be achieved at the narrowest welded width of 2.5 mm, with retention of more than 99% tensile strength. The FEM result indicated that welded joints could be obtained with strength that was comparable to that of the base material with a narrow softening region, even when using an undermatched filler wire. However, as shown by the result in Figure 21, the softer weld metal joint failed at the weld metal region, raising concerns about a decrease in fracture elongation and other properties.

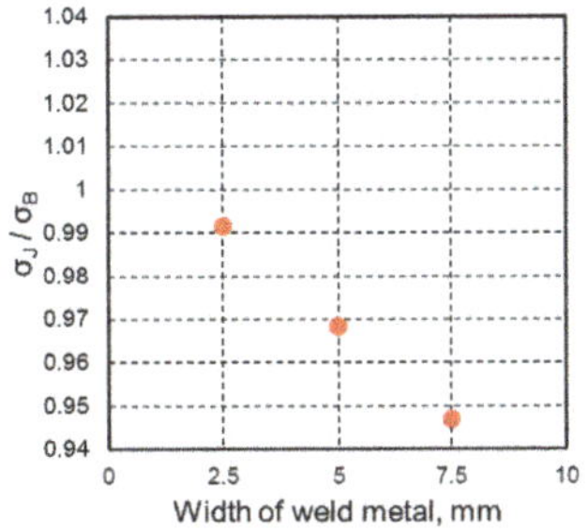


**Figure 22.** Effect of weld metal width on joint efficiency ($\sigma_J/\sigma_B$).

To confirm the effect of the softened HAZ width on joint strength, analyses were carried out at three positions of the softened zone width in an even-matched joint, which is defined as constrained by equivalent hardnesses of the weld metal and base metal. The analyses were carried out on the basis of the conditions shown in Table 5. Contour diagrams of the mean normal stress at the time of yielding, at the start of plastic deformation (displacement: 0.415 mm), and at the end of deformation are shown in Figure 23. At the start of deformation, there was no stress concentration in the weld metal and HAZ regions, and similarity in all conditions. Deformation progress to softened zone widths of 1.0 mm and 2.0 mm clearly showed a low-stress (blue) region adjacent to the weld metal, which is a high-hardness region. At the end of deformation, the 1.0 mm softened zone width showed large deformation and thickness decrease in the base metal, and the highest stress and necking deformation occurred approximately 10 mm from the center of the weld metal. In contrast, 1.5 mm and 2.0 mm softened zone widths showed the highest stress and necking deformation at the softened HAZ region.

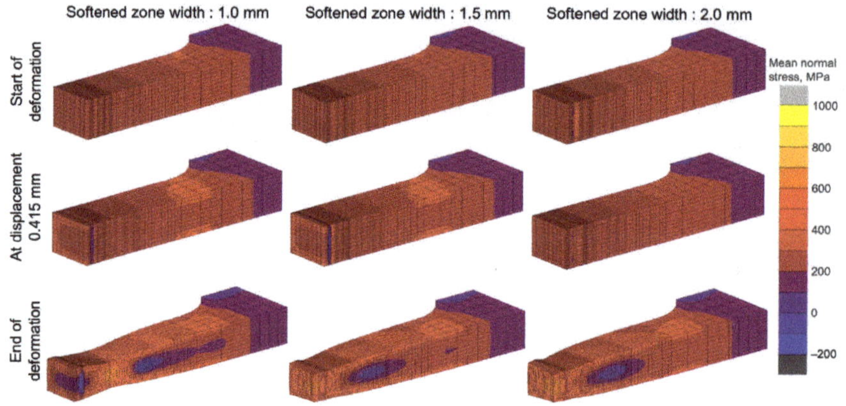


**Figure 23.** Mean nominal stress contours and deformation in analysis of softened heat-affected zone width effect.

The effect of joint efficiency on softened zone width was summarized by considering the nominal stress of the joint ($\sigma_J$) during FEM analysis of tensile strength. The result was calculated with nominal tensile stress of the base material ($\sigma_B$) and is represented in Figure 24. The results indicate that with the narrowest softened zone (1.0 mm), the joint efficiency reached 100% compared with that of the base material. In contrast, a wider softened zone resulted in lower joint efficiency due to plastic deformation in the softened region, leading to lower strength. On the basis of these findings, several articles [47–49] have reported that an increase in strength and decrease in the width of the softened HAZ can enhance the strength of the butt joints: effects of the strength and width of a hardened HAZ on butt joint strength were insignificant.

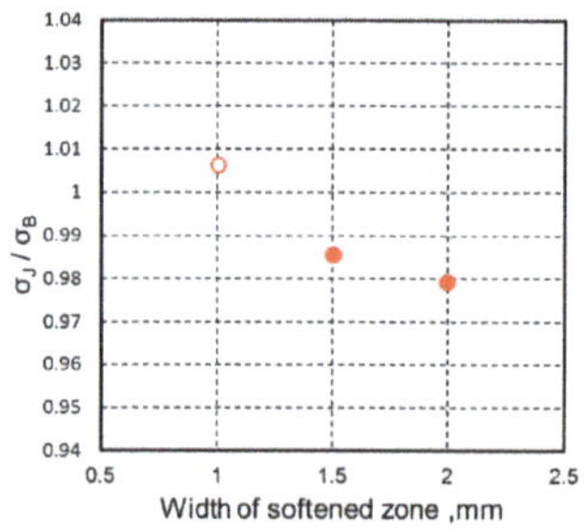


**Figure 24.** Effect of softened heat-affected zone width on joint efficiency ($\sigma_J/\sigma_B$) (red dot: < 1.0, red circle: > 1.0).

Failure locations can occur in either the weld metal or base metal, depending on the mismatch ratio and width of the softened zone. Mechanical behavior of the overall joint strength is influenced by the plastic distribution within the softened HAZ. Narrow softened regions can achieve strengths comparable to those of the welded metal or base metal due to the constraints imposed by the adjacent unyielded materials. In cases of mismatched joints, a decrease in joint ductility is observed, which varies based on the failure location and hardening properties in the failure zone [49]. However, on the basis of our FEM results, achieving a highly efficient joint requires a strategy to narrow the HAZ soft zone as much as possible. This approach minimizes plastic deformation in the softened region, thereby reducing deformation and stress on the base metal and ensuring that the ductility remains comparable to that of the base metal.

## 4. Conclusions

The effect of welding speed on the microstructure and hardness of a welded joint was investigated under conditions of welding 10 mm thick 780 MPa-class high-tensile strength steel in a single pass using hot-wire laser narrow-gap welding. The heat input was estimated from the measured grain size in the HAZ. In addition, on the basis of the experimental results, a strength-evaluation specimen was welded under optimized welding conditions. The effects of undermatched weld metal width and softened HAZ width on the joint strength were clarified using FEM analysis. The conclusions obtained in this study are as follows.

1. Welding at various speeds of 0.3, 0.5, and 0.75 m/min, and hot-wire narrow-gap laser welding was performed in a single pass using 490 MPa-class wire as the filler wire and 10 mm thick 780 MPa-class high-tensile steel. The lack-of-fusion defect ratio was minimized at welding speeds of 0.3 and 0.5 m/min, and no other defects occurred. There was a difference in the hardness of the weld metal, and the weld speed of 0.3 m/min resulted in an undermatched joint due to a slower cooling speed compared with those welded at 0.5 and 0.75 m/min.
2. The HAZ structures were observed. Grain sizes in the coarsening region near the weld metal were approximately 100 μm, 60 μm, and 40 μm at welding speeds of 0.3 m/min, 0.5 m/min, and 0.75 m/min, respectively. On the basis of these results, the respective heat inputs were estimated as 11.0 kJ/cm, 7.5 kJ/cm, and 4.5 kJ/cm. The estimated heat inputs are considered significantly low compared with those of other welding methods. This confirmed that a low-heat input hot-wire laser welding process had been carried out.
3. Charpy impact and tensile tests were carried out on joints that had been welded under optimized conditions. Charpy impact tests were carried out on the weld metal and at the fusion boundary. It was confirmed that the absorbed energy was 47 J or more at all positions when the test temperature was −20 °C. The location of the tensile test failure in the base metal confirmed that it was possible to obtain sound strength, compared with that of the base metal, even if softening occurred in the HAZ.
4. FEM analysis was carried out for various widths of the softened area to confirm the effect of the weld metal on the strength of an undermatched joint. When the width of the weld metal was limited to 2.5 mm, the joint efficiency was 99% or more and, even when a soft weld metal area was produced, it was possible to obtain a tensile strength almost the same as that of the base material.
5. FEM analysis was carried out to confirm the effect of changes in the HAZ softening region width on the strength of the joint. It was confirmed that the joint efficiency reached 100% when the softened HAZ width was very narrow (1.0 mm in this study), and fracture occurred in the base metal.
6. Use of the narrow-gap hot-wire laser welding method enabled production of sound joints with a high balance between strength and toughness by the suppression of HAZ softening using a low-heat input process and the application of soft wire.

In the next study, investigations of other kinds of base materials and filler wires to achieve sound joints with higher balance of some properties will be performed. In addition, the hot-wire laser welding process will be applied to enable more efficient welding for thicker plates with sufficient joint properties.

**Author Contributions:** Conceptualization, J.G., K.M. (Kazuyuki Matsumoto), K.M. (Keita Marumoto) and M.Y.; methodology, J.G. and K.M. (Keita Marumoto); validation, K.M. (Kazuyuki Matsumoto) and K.M. (Keita Marumoto); formal analysis, J.G. and K.M. (Keita Marumoto); investigation, J.G. and K.M. (Keita Marumoto); resources, K.M. (Kazuyuki Matsumoto) and M.Y.; data curation, J.G. and K.M. (Keita Marumoto); writing—original draft preparation, J.G., K.M. (Kazuyuki Matsumoto) and K.M. (Keita Marumoto); writing—review and editing, K.M. (Keita Marumoto) and M.Y.; visualization, K.M. (Kazuyuki Matsumoto) and K.M. (Keita Marumoto); supervision, M.Y.; project administration, M.Y.; funding acquisition, K.M. (Kazuyuki Matsumoto) and M.Y. All authors have read and agreed to the published version of the manuscript.

**Funding:** This research received no external funding.

**Institutional Review Board Statement:** Not applicable.

**Informed Consent Statement:** Not applicable.

**Data Availability Statement:** The original contributions presented in the study are included in the article; further inquiries can be directed to the corresponding author.

**Acknowledgments:** Daik Okita and Yuta Mashino provided valuable materials for this study, which is acknowledged and appreciated.

**Conflicts of Interest:** The authors declare no conflicts of interest.

## References

1. Marumoto, K.; Fujinaga, A.; Takahashi, T.; Yamamoto, H.; Yamamoto, M. Selection of Welding Conditions for Achieving Both High Efficiency and Low Heat Input for Hot-Wire Gas Metal Arc Welding. *JMMP* **2024**, *8*, 82. [CrossRef]
2. Khan, M.S.; Soleimani, M.; Midawi, A.R.H.; Aderibigbe, I.; Zhou, Y.N.; Biro, E. A Review on Heat Affected Zone Softening Dual-Phase Steels During Laser Welding. *J. Manuf. Process.* **2023**, *102*, 663–684. [CrossRef]
3. Chen, L.; Nie, P.; Qu, Z.; Ojo, O.A.; Xia, L.; Li, Z.; Huang, J. Influence of Heat Input on the Changes in the Microstructure an Fracture Behavior of Laser Welded 800MPa Grade High-Strength Low-Alloy Steel. *J. Manuf. Process.* **2020**, *50*, 132–141. [CrossRe
4. Gáspár, M. Effect of Welding Heat Input on Simulated HAZ Areas in S960QL High Strength Steel. *Metals* **2019**, *9*, 1226. [CrossRe
5. Ramakrishna, R.V.S.M.; Amrutha, P.H.S.L.R.; Rahman Rashid, R.A.; Palanisamy, S. Narrow Gap Laser Welding (NGLW) Structural Steels-a Technological Review and Future Research Recommendations. *Int. J. Adv. Manuf. Technol.* **2020**, *111*, 2277–23 [CrossRef]
6. Yang, W.; Xin, J.; Fang, C.; Dai, W.; Wei, J.; Wu, J.; Song, Y. Microstructure and Mechanical Properties of Ultra-Narrow Gap Las Weld Joint of 100 Mm-Thick SUS304 Steel Plates. *J. Mater. Process. Technol.* **2019**, *265*, 130–137. [CrossRef]
7. Ning, J.; Zhang, L.-J.; Yang, J.; Yin, X.-Q.; Wang, X.-W.; Wu, J. Characteristics of Multi-Pass Narrow-Gap Laser Welding of D40 Ultra-High Strength Steel. *J. Mater. Process. Technol.* **2019**, *270*, 168–181. [CrossRef]
8. Vara Prasad, V.; Madhu Babu, C.; Ajay, P. A Review On Rotating Arc Welding Process. *Mater. Today Proc.* **2018**, *5*, 3551–35 [CrossRef]
9. Zhao, Y.; Ma, S.; Huang, J.; Wu, Y. Narrow-Gap Laser Welding Using Filler Wire of Thick Steel Plates. *Int. J. Adv. Manuf. Techn* **2017**, *93*, 2955–2962. [CrossRef]
10. Li, R.; Wang, T.; Wang, C.; Yan, F.; Shao, X.; Hu, X.; Li, J. A Study of Narrow Gap Laser Welding for Thick Plates Using t Multi-Layer and Multi-Pass Method. *Opt. Laser Technol.* **2014**, *64*, 172–183. [CrossRef]
11. Wang, J.; Zhu, J.; Fu, P.; Su, R.; Han, W.; Yang, F. A Swing Arc System for Narrow Gap GMA Welding. *ISIJ Int.* **2012**, *52*, 110–1 [CrossRef]
12. Suwannatee, N.; Yamamoto, M.; Shinohara, S. Optimization of Hot-Wire Fraction for Enhanced Quality in GMAW. *Weld. Wor* **2023**, *68*, 1017–1032. [CrossRef]
13. Budde, L.; Biester, K.; Lammers, M.; Hermsdorf, J.; Kaierle, S.; Overmeyer, L. Influence of Process Parameters on Single We Seam Geometry and Process Stability in Laser Hot-Wire Cladding of AISI 52100. *Adv. Ind. Manuf. Eng.* **2023**, *7*, 100122. [CrossRe
14. Li, J.; Sun, Q.; Kang, K.; Zhen, Z.; Liu, Y.; Feng, J. Process Stability and Parameters Optimization of Narrow-Gap Laser Vertic Welding with Hot Wire for Thick Stainless Steel in Nuclear Power Plant. *Opt. Laser Technol.* **2020**, *123*, 105921. [CrossRef]
15. Todo, M.; Shinozaki, K.; Yamamoto, M.; Kadoi, K.; Yamamoto, M.; Phaonaim, R.; Okagaito, T. Hot-Wire Laser Welding Proce Using Laser Diode for Large-Diameter Pipe with Narrow Gap Joint. *Q. J. Jpn. Weld. Soc.* **2015**, *33*, 107s–110s. [CrossRef]

Phaoniam, R.; Shinozaki, K.; Yamamoto, M.; Kadoi, K.; Tsuchiya, S.; Nishijima, A. Development of a Highly Efficient Hot-Wire Laser Hybrid Process for Narrow-Gap Welding-Welding Phenomena and Their Adequate Conditions. *Weld World* **2013**, *57*, 607–613. [CrossRef]
Jiang, J.; Peng, Z.Y.; Ye, M.; Wang, Y.B.; Wang, X.; Bao, W. Thermal Effect of Welding on Mechanical Behavior of High-Strength Steel. *J. Mater. Civ. Eng.* **2021**, *33*, 04021186. [CrossRef]
Chen, C.; Chiew, S.-P.; Zhao, M.-S.; Lee, C.-K.; Fung, T.-C. Influence of Cooling Rate on Tensile Behaviour of S690Q High Strength Steel Butt Joint. *J. Constr. Steel Res.* **2020**, *173*, 106258. [CrossRef]
Schroepfer, D.; Witte, J.; Kromm, A.; Kannengiesser, T. Stresses in Repair Welding of High-Strength Steels-Part 1: Restraint and Cold Cracking Risk. *Weld. World* **2024**, *68*, 685–697. [CrossRef]
Xin, H.; Correia, J.A.F.O.; Veljkovic, M.; Berto, F.; Manuel, L. Residual Stress Effects on Fatigue Life Prediction Using Hardness Measurements for Butt-Welded Joints Made of High Strength Steels. *Int. J. Fatigue* **2021**, *147*, 106175. [CrossRef]
John, M.; Kumar, P.A.; Bhat, K.U. Effect of Filler Wire Strength on High Strength Low Alloy Steels. *Mater. Today Proc.* **2022**, *49*, 1286–1293. [CrossRef]
Kim, D.-Y.; Lee, T.H.; Kim, C.; Kang, M.; Park, J. Gas Metal Arc Welding with Undermatched Filler Wire for Hot-Press-Formed Steel of 2.0 GPa Strength: Influence of Filler Wire Strength and Bead Geometry. *Mater. Today Commun.* **2023**, *34*, 105244. [CrossRef]
Hochhauser, F.; Ernst, W.; Rauch, R.; Vallant, R.; Enzinger, N. Influence of the Soft Zone on the Strength of Welded Modern HSLA Steels. *Weld. World* **2012**, *56*, 77–85. [CrossRef]
Rodrigues, D.M.; Menezes, L.F.; Loureiro, A.; Fernandes, J.V. Numerical Study of the Plastic Behaviour in Tension of Welds in High Strength Steels. *Int. J. Plast.* **2004**, *20*, 1–18. [CrossRef]
Ikawa, H.; Oshige, H.; Noi, S.; Date, H.; Uchikawa, K. Study on the Grain Growth in Weld-Heat Affected Zone (Report 7). *J. Jpn. Weld. Soc.* **1997**, *46*, 636–641. [CrossRef]
*JIS Z2241*; Metallic materials-Tensile testing-Method of test at room temperature. Japanese Industrial Standards: Tokyo, Japan, 2022.
Pavlina, E.J.; Van Tyne, C.J. Correlation of Yield Strength and Tensile Strength with Hardness for Steels. *J. Mater. Eng. Perform.* **2008**, *17*, 888–893. [CrossRef]
Han, Q.; Gao, J.; Han, C.; Zhang, G.; Li, Y. Experimental Investigation on Improving the Deposition Rate of Gas Metal Arc-Based Additive Manufacturing by Auxiliary Wire Feeding Method. *Weld. World* **2020**, *65*, 35–45. [CrossRef]
Shehryar Khan, M.; Shahabad, S.I.; Yavuz, M.; Duley, W.W.; Biro, E.; Zhou, Y. Numerical Modelling and Experimental Validation of the Effect of Laser Beam Defocusing on Process Optimization During Fiber Laser Welding of Automotive Press-Hardened Steels. *J. Manuf. Process.* **2021**, *67*, 535–544. [CrossRef]
Hung, C.-H.; Chen, W.-T.; Sehhat, M.H.; Leu, M.C. The Effect of Laser Welding Modes on Mechanical Properties and Microstructure of 304L Stainless Steel Parts Fabricated by Laser-Foil-Printing Additive Manufacturing. *Int. J. Adv. Manuf. Technol.* **2020**, *112*, 867–877. [CrossRef]
Fu, Y.; Guo, N.; Cheng, Q.; Zhang, D.; Feng, J. Underwater Laser Welding for 304 Stainless Steel with Filler Wire. *J. Mater. Res. Technol.* **2020**, *9*, 15648–15661. [CrossRef]
Tümer, M.; Schneider-Bröskamp, C.; Enzinger, N. Fusion Welding of Ultra-High Strength Structural Steels-A Review. *J. Manuf. Process.* **2022**, *82*, 203–229. [CrossRef]
Bao, L.; Wang, Y.; Han, T. Study on Microstructure-Toughness Relationship in Heat Affected Zone of EQ70 Steel by Laser-Arc Hybrid Welding. *Mater. Charact.* **2021**, *171*, 110788. [CrossRef]
Afkhami, S.; Javaheri, V.; Amraei, M.; Skriko, T.; Piili, H.; Zhao, X.-L.; Björk, T. Thermomechanical Simulation of the Heat-Affected Zones in Welded Ultra-High Strength Steels: Microstructure and Mechanical Properties. *Mater. Des.* **2022**, *213*, 110336. [CrossRef]
Li, X.; Ma, X.; Subramanian, S.V.; Shang, C.; Misra, R.D.K. Influence of Prior Austenite Grain Size on Martensite-Austenite Constituent and Toughness in the Heat Affected Zone of 700MPa High Strength Linepipe Steel. *Mater. Sci. Eng. A* **2014**, *616*, 141–147. [CrossRef]
Chen, S.; Tsang, D.K.L.; Jiang, L.; Yu, K.; Li, C.; Li, Z.; Zhou, X.; Yang, J. Microstructure and Local Strains in GH3535 Alloy Heat Affected Zone and Their Influence on the Mechanical Properties. *Mater. Sci. Eng. A* **2017**, *699*, 48–54. [CrossRef]
Tümer, M.; Vallant, R.; Warchomicka, F.G.; Enzinger, N. Undermatched Welding of Ultra-High-Strength Steel S1100 with Metal-Cored Wire: Influence of Welding Positions on Mechanical Properties. *J. Mater. Eng. Perform.* **2022**, *31*, 7068–7079. [CrossRef]
Ye, Z.; Chen, G.; Wang, W.; Wang, S.; Yang, J.; Huang, J. High-Strength and High-Toughness Weld of S690QL HSLA Steel via a High-Speed High Frequency Electric Cooperated Arc Welding Without Additional Heating Measures. *Mater. Sci. Eng. A* **2023**, *880*, 145328. [CrossRef]
Gao, S.; Yue, Y.; Zhang, D.; Li, N.; Huang, Y.; Yan, J.; Yang, L. Effect of heat input on the mechanical properties of welded Q960E ultra-high strength steel butt joints. *Eng. Struct.* **2024**, *303*, 117503. [CrossRef]
Liu, X.; Chung, K.-F.; Ho, H.-C.; Xiao, M.; Hou, Z.-X.; Nethercot, D.A. Mechanical behavior of high strength S690-QT steel welded sections with various heat input energy. *Eng. Struct.* **2018**, *175*, 245–256. [CrossRef]

41. Yang, X.; Di, X.; Liu, X.; Wang, D.; Li, C. Effects of heat input on microstructure and fracture toughness of simulated coarse-grain heat affected zone for HSLA steels. *Mater. Charact.* **2019**, *155*, 109818. [CrossRef]
42. Wen, C.; Wang, Z.; Deng, X.; Wang, G.; Misra, R.D.K. Effect of Heat Input on the Microstructure and Mechanical Properties Low Alloy Ultra-High Strength Structural Steel Welded Joint. *Steel Res. Int.* **2018**, *89*, 1700500. [CrossRef]
43. Guo, W.; Li, L.; Crowther, D.; Dong, S.; Francis, J.A.; Thompson, A. Laser welding of high strength steels (S960 and S700) wi medium thickness. *J. Laser Appl.* **2016**, *28*, 022425. [CrossRef]
44. Zhang, H.; Jiang, M.; Chen, X.; Wei, L.; Wang, S.; Jiang, Y.; Jiang, N.; Wang, Z.; Lei, Z.; Chen, Y. Investigation of Weld Root Defec in High-Power Full-Penetration Laser Welding of High-Strength Steel. *Materials* **2022**, *15*, 1095. [CrossRef] [PubMed]
45. Schneider, C.; Ernst, W.; Schnitzer, R.; Staufer, H.; Vallant, R.; Enzinger, N. Welding of S960MC with undermatching filler materi *Weld. World* **2018**, *62*, 801–809. [CrossRef]
46. Lee, Y.; Park, S.; Seon, J.; Han, H.; Lee, K.; Kang, N. Overcoming underfill defect with undermatching filler and establishir new acceptance criteria for underfill depth in laser beam welding of ultra-high strength steels. *Weld. World* **2023**, *68*, 259–2 [CrossRef]
47. Ran, M.-M.; Sun, F.-F.; Li, G.-Q.; Wang, Y.-B. Mechanical properties of mismatched high strength steel butt joints with thr softened/hardened strength distribution patterns. *Thin-Walled Struct.* **2020**, *146*, 106456. [CrossRef]
48. Ran, M.-M.; Sun, F.-F.; Li, G.-Q.; Kanvinde, A.; Wang, Y.-B.; Xiao, R.Y. Experimental study on the behavior of mismatched bu welded joints of high strength steel. *J. Constr. Steel Res.* **2019**, *153*, 196–208. [CrossRef]
49. Cai, W.-Y.; Wang, Y.-B.; Li, G.-Q.; Stroetmann, R. Comparative study on strength of TMCP and QT high-strength steel butt-weld joints. *J. Constr. Steel Res.* **2022**, *197*, 107447. [CrossRef]

materials

MDPI

*icle

# esearch on Laser Cleaning of Graphite Lubrication Coating on ıe Magnesium Alloy Surface

enhai Xu [1,2,3], Yunhui Yue [3], Donghe Zhang [2], Shaoxi Xue [3], Erju Liu [3], Debin Shan [2,3,*], Jie Xu [2,3] and Bin Guo [2,3]

[1] CGN-HIT Advanced Nuclear and New Energy Research Institute, Harbin Institute of Technology, Harbin 150001, China; zhenhaixu@hit.edu.cn
[2] Zhengzhou Research Institute, Harbin Institute of Technology, Zhengzhou 450000, China; zhangdonghe@hit.edu.cn (D.Z.); xjhit@hit.edu.cn (J.X.); bguo@hit.edu.cn (B.G.)
[3] National Key Laboratory for Precision Hot Processing of Metals, Harbin Institute of Technology, Harbin 150001, China; 22s009106@stu.hit.edu.cn (Y.Y.); xueshaoxi@hit.edu.cn (S.X.); 20b909103@stu.hit.edu.cn (E.L.)
* Correspondence: shandb@hit.edu.cn

**Abstract:** The lubricating coating must be removed from the forged or stamped workpieces. Developing environment-friendly and high-precision cleaning technology is necessary. In this study, a nanosecond pulsed laser was used to clean the graphite lubricating coating of 15 μm thickness on the surface of an MB15 magnesium alloy. The effects of various laser cleaning parameters on the cleaning quality and the cleaning mechanism were studied. When the laser fluence ($F$) increases from 1.27 to 7.64 J/cm$^2$, the clearance rate increases, and the surface roughness initially decreases before increasing. When the pulse frequency ($f$) increases from 10 to 30 kHz, the single-pulse energy decreases, the clearance rate decreases, and the surface roughness increases. When the scanning speed ($v$) increases from 1000 to 5000 mm/s, the spot overlap rate decreases, the clearance rate decreases, and the surface roughness firstly decreases and then increases. The optimal cleaning parameter combinations are $F$ = 3.82 J/cm$^2$, $f$ = 10 kHz, and $v$ = 3000 mm/s. The graphite lubrication coating was almost completely removed without damaging the substrate surface, and the surface carbon content of the sample was decreased to 6.42%. The laser cleaning mechanism of the graphite lubricating coating on the magnesium alloy surface is dominated by thermal ablation. As the laser fluence increases, the physical and chemical reactions become more violent.

**Keywords:** laser cleaning; graphite lubrication coating; removal mechanism; surface morphology; roughness; magnesium alloy

## 1. Introduction

During the forging or stamping process, the friction between the workpiece and the mold can lead to the difficulty of forming, the increased wear, and the premature failure of the mold. In order to enhance the product quality, minimize the wear, and prolong the mold life, utilizing high-efficacy lubricants like graphite and molybdenum disulfide is a common practice. However, after forming, they often adhere to the surface of the workpiece and need to be removed further. The traditional chemical cleaning method [1–4] utilizes chemical reactions to decrease the adhesion between dirt and the substrate. However, the prolonged use of chemical agents can corrode the substrate surface, leading to environmental pollution from the resulting waste liquid. The mechanical cleaning method [5–7] utilizes the friction and shear effects of abrasives on the dirt coating for removal. However, it is limited in cleaning complex-shaped parts, and the cleaning accuracy is uncontrollable, making automation challenging. The ultrasonic cleaning method [8,9] cannot effectively clean submicron particles due to its inherent limitations. Against the backdrop of vigorously promoting the transformation and upgrading of the manufacturing industry globally, traditional cleaning methods can no longer meet the high-precision and pollution-free requirements of various fields for cleaning processes. Based on this, finding an efficient method to remove the lubricant adhered to the surface of the workpiece in an environmentally friendly manner without damaging the substrate is a novel concept for the surface cleaning of forgings.

Laser cleaning, as a new type of cleaning technology, is considered the most promising green cleaning technology of the 21st century [10–13]. It has the advantages of controllability, efficacy, wide applicability, non-contact operation, and environmentally friendly green protection. In 1969, the concept of laser cleaning was initially proposed by American scientists S. M. Bedair and Harold P. Smith, Jr. They successfully removed the oxygen or sulfur contamination from the surface of nickel by exposing it to high-power laser pulses (120 $MW/cm^2$). This laser-based cleaning method has served as a source of inspiration for researchers, leading to extensive and profound investigations into laser technology and its underlying mechanisms [14].

At present, laser cleaning technology has been applied in metal surface treatment (removal of oxide film [15–17], paint [18–20], rust [21,22]), semiconductor industry [23–25], and for cultural relic protection [26–28]. Zhang et al. [29] used a laser to remove the oxide film from the surface of 5754 aluminum alloy and found that the oxide coating could be entirely eliminated by the laser. The researchers explained the cleaning mechanism and plasma behavior by comparing the cleaning efficacy on aluminum alloys at various energy densities. Zhang et al. [30] used a nanosecond pulsed laser to remove the blue paint coating on an Al-Mg aluminum alloy substrate and found that the thickness of the removed coating gradually increased with low laser irradiation. The main mechanisms of blue coating detachment from the surface are ablation, peeling, and vaporization. Wang et al. [31] conducted laser cleaning experiments on the surface rust of steel using pulsed Nd: YAG and found that under appropriate laser parameters, the pulsed Nd: YAG laser can completely remove the rust coating. Moreover, the microhardness and corrosion resistance of the surface improved after laser cleaning, with the enhancement correlating with the increase in laser energy. Ye et al. [32] used Nd: YAG laser to induce optical breakdown of the airborne above the gold-coated K9 glass surface, and the created shockwave removed the SiO2 particles contaminated on the gold films. The results showed that the 1064 nm laser-induced plasma shockwave can effectively remove the SiO2 particles; the removal ratio can reach above 90%. In summary, laser cleaning can achieve high-quality cleaning efficacy for various objects without damaging the substrate by utilizing a suitable combination of laser parameters. For a specific cleaning task, the influence of laser parameters on contaminant removal can be investigated by analyzing experimental data from laser cleaning, optimizing process parameters, and understanding the characteristics of laser cleaning. Therefore, laser cleaning technology provides a new option for removing lubricating coatings from the surface of formed specimens.

So far, there are few reports on the use of laser cleaning technology for removing lubricating coatings on the surface of formed specimens. Therefore, exploring this research is of great significance. This study utilized a nanosecond pulsed laser cleaning system to conduct experiments on removing the graphite lubrication coating from the surface of MB15 magnesium alloy. The study investigated the impact of laser fluence, pulse frequency, and scanning speed on the cleaning efficacy. This study evaluated the surface characteristics of the cleaned samples using optical microscopy, scanning electron microscopy (SEM), energy dispersive spectroscopy (EDS), and laser confocal microscopy. It also explored the mechanism of laser cleaning of the graphite coating by employing a high-speed camera to observe the dynamic behavior and physical phenomena during the laser cleaning process. This study addresses the gap in laser cleaning of lubricating coatings on forgings and provides a theoretical foundation for the subsequent application of laser cleaning on the surface lubricating coatings of formed workpieces.

## 2. Materials and Methods

### 2.1. Materials and Samples Preparation

In this experiment, MB15 magnesium alloy samples with dimensions of 20 mm × 20 mm × 4 mm (length × width × thickness) were prepared, and their main elemental composition is shown in Table 1. The samples were meticulously polished using sandpaper and thoroughly cleaned with alcohol to ensure surface cleanliness. The samples were heated in a furnace to around 110 °C, then water-based graphite emulsion was sprayed on them. After the water in the atomized graphite evaporated, the graphite adhered to the surface of the magnesium alloy, forming the graphite lubricating coating of 15 μm. The macro and micromorphology of the sample surface and coating thickness are shown in Figure 1.

**Table 1.** MB15 magnesium alloy chemical composition (wt%).

| Al | Mn | Zn | Zr | Cu | Ni | Si | Fe | Be | Mg |
|---|---|---|---|---|---|---|---|---|---|
| ≤0.050 | ≤0.100 | 5.000~6.000 | 0.300~0.900 | ≤0.050 | ≤0.005 | ≤0.050 | ≤0.050 | ≤0.010 | Bal |

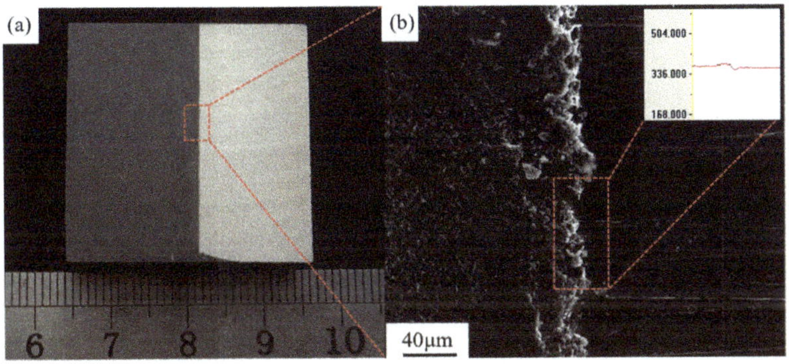


**Figure 1.** Surface morphology of the sample with the graphite lubricating coating at: (**a**) Macro level; (**b**) Microscopic level; The inset shows the measure result of the coating thickness using a confocal laser microscope.

### 2.2. Laser Cleaning Experimental Equipment

The pulsed laser cleaning device used in this experiment is shown in Figure 2a, which consists of five parts: nanosecond fiber laser, scanning galvanometer, control system, industrial robot, and dust removal system. Its schematic diagram is shown in Figure 2b. A nanosecond pulsed fiber laser (YLPN-100-30×100-1000, IPG photonics, Marlborough, MA, USA) with a maximum average power of 1000 W was employed in the experiments, which operated at a laser wavelength of 1064 nm, a spot diameter of 1 mm, and a maximum laser scanning range of 100 mm × 100 mm. The detailed laser cleaning parameters are listed in Table 2. Experimental samples were fixed on an X-Y horizontal platform, and the position of the light output can be adjusted to align with the sample using a robotic arm. Laser light was transmitted through optical fibers and directed by a two-dimensional galvanometer scanning head to conduct Z-shaped reciprocating scanning on the X-Y plane, aiming to clean the specified area of the target plane.

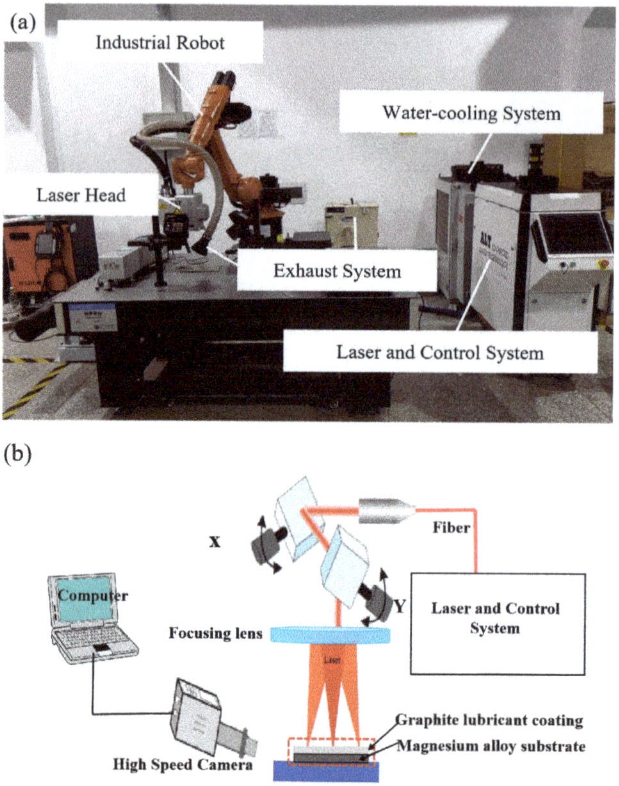


**Figure 2.** (**a**) Laser cleaning device; (**b**) Schematic diagram of laser cleaning device principle.

**Table 2.** Laser cleaning parameters.

| Parameter | Symbol | Value | Unit |
|---|---|---|---|
| Wavelength | $\lambda$ | 1064 | nm |
| Average laser power | $p$ | <1000 | W |
| Spot diameter | $D$ | 1 | mm |
| Pulsed width | $\tau$ | 30, 40, 60, 100 | ns |
| Pulse frequency | $f$ | $\leq$50 | kHz |
| scanning speed | $v$ | $\leq$5000 | mm/s |
| Overlap rate | $O$ | 50–90 | % |

*2.3. Characterization Methods*

Metallographic microscopy (HAL-100, Carl Zeiss AG, Oberkochen, Germany) and scanning electron microscopy (Quanta 200FEG, FEI, Hillsborough, OR, USA and SU5000, HITACHI, Tokyo, Japan) were used to observe the microstructure of the cleaned sample surface. X-ray energy-dispersive spectroscopy (EDS, ULTIMATELY MAX40, Oxford Instruments, Oxford, UK) was used to analyze the elemental content of the cleaned sample surface. A laser scanning confocal microscope (OLS-3000, HITACHI OLYMPUS, Tokyo, Japan) was used to measure the surface roughness of the cleaned sample and observe the 3D surface morphology. High-speed cameras (FASTCAM Mini UX 100, PHOTRON, Tokyo, Japan) were utilized to monitor in real time the dynamic behavior of the decomposition and peeling of lubricating coatings on the substrate surface during laser cleaning.

## 3. Results

### 3.1. Laser Cleaning Process and Mechanism

When a pulsed laser irradiates the surface of the material to be cleaned, it interacts with the material. Part of the laser that strikes the sample is reflected and transmitted, while the rest is absorbed by the sample surface. At the microscopic level, the laser energy is absorbed by electrons, atoms, and lattice vibrations, leading to changes in energy. At the macroscopic level, this manifests as surface material gasification, combustion, melting, and explosive debris splashing, all of which are accompanied by the removal of surface contaminants. This study used a high-speed camera to capture real-time dynamic behavior during laser cleaning and recorded the dynamic process of pulsed laser acting on the surface of the sample.

During the laser cleaning process, the light spot is output to the surface to be cleaned. Figure 3 shows the process of the laser spot walking on the surface of the sample. In the image, the area above the red line annotation is the uncleaned area, while the area below the red line represents the cleaned area. Comparing between Figure 3a,c, it can be observed that the laser has a significant cleaning effect on the area it has traveled through. Compared to the already cleaned area in Figure 3a, the area in Figure 3c that has been completely cleaned has increased. Figure 3b shows the phenomenon of the light spot walking, and it can be clearly seen in the figure that the area scanned to the left by the light spot has a significant cleaning effect, exposing a bright magnesium alloy substrate. Based on the laser being used as a reference, comparing the cleaned area on the right and the uncleaned area on the left, it can be concluded that laser cleaning can effectively remove the graphite lubricating film coating on the surface of the magnesium alloy.

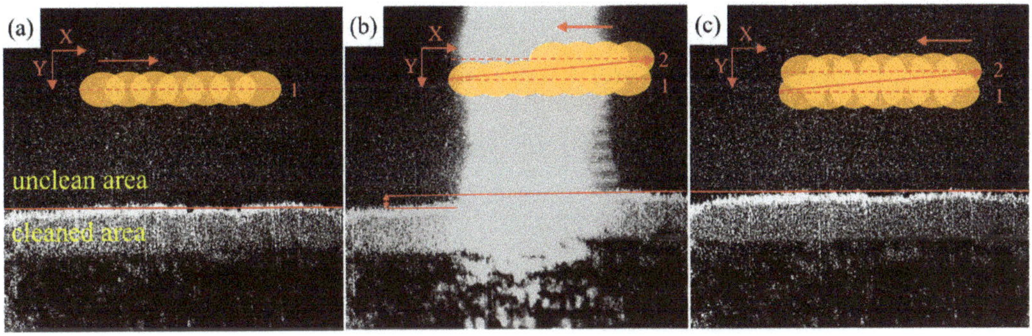


**Figure 3.** The process of laser cleaning: (**a**) laser spot completes scanning path 1; (**b**) laser spot is moving; (**c**) laser spot completes scanning path 2. The dashed lines indicate the laser scanning paths, and the arrows indicate the scanning directions.

At present, the interaction mechanism between laser and materials can be divided into photothermal effect, photochemical effect, and photophysical effect. Usually, the removal of materials is caused by the combined action of several mechanisms. This study selected three representative cleaning processes—namely, partial removal (1.27 J/cm$^2$), complete removal (3.82 J/cm$^2$), and substrate damage (6.37 J/cm$^2$)—for high-speed camera observation to analyze the dynamic behavior of coating peeling and reveal the mechanism of laser cleaning of graphite lubricating coating on the magnesium alloy surface. The laser is irradiated onto the surface to be cleaned, causing the coating to absorb energy, increase in temperature, and undergo thermal conduction. As the coating temperature rises to a certain value, a series of physical and chemical reactions such as melting, evaporation, and combustion will occur. Large blocks of material are transformed into small blocks or molecules, causing the coating to detach from the substrate surface. The increase in temperature is positively correlated with the laser fluence, and the larger the laser fluence the higher the instantaneous temperature of the coating, and the more severe the removal phenomenon.

As shown in Figure 4, as the laser fluence increases, the laser cleaning effect improves and the reaction becomes more violent [33,34]. During the cleaning process captured by high-speed cameras, obvious thermal ablation behavior can be observed mainly manifested in two forms: selective evaporation and peeling of the evaporation reaction. After the laser spot moves, a large amount of smoke, dust, and debris can be observed. And as the laser fluence increases, the combustion becomes more intense and the amount of debris splashing around increases. The coating absorbs laser energy, causing various physical and chemical changes, including gasification, ablation, and peeling, to remove the graphite coating together [30,35,36]. When the laser fluence is 1.27 J/cm$^2$, although the cleaning threshold of graphite coating is reached, the energy is not enough, and only a part of the graphite film coating is removed. As the laser energy increases, the graphite coating absorbs more energy, the physical and chemical reactions become more violent, and the graphite film coating is completely removed.

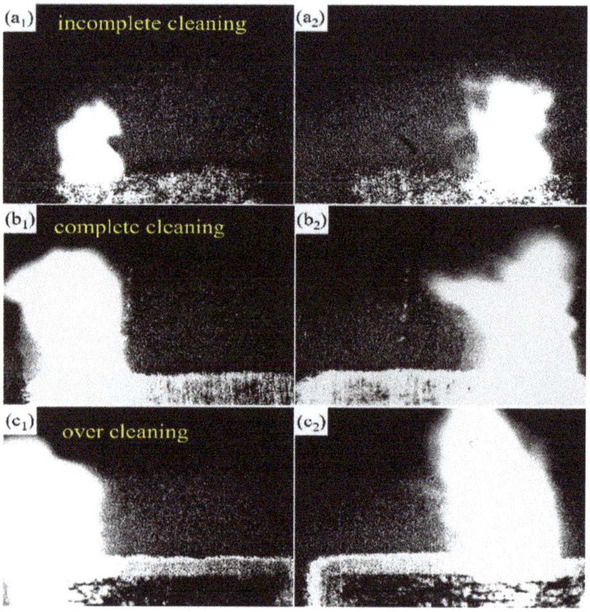


**Figure 4.** Laser cleaning dynamic behavior: ($\mathbf{a_1}$,$\mathbf{a_2}$) 1.27 J/cm$^2$; ($\mathbf{b_1}$,$\mathbf{b_2}$) 3.82 J/cm$^2$; ($\mathbf{c_1}$,$\mathbf{c_2}$) 6.37 J/cm$^2$.

### 3.2. *The Effect of the Laser Fluence on Cleaning*

The above analysis of the laser cleaning process has concluded that laser cleaning technology can effectively remove graphite lubricating coatings on the surface of magnesium alloys. Laser cleaning technology has application and research value for removing the lubricating coating on the surface of hot formed workpieces.

To study the laser cleaning characteristics of graphite lubricating coatings and determine the optimal process parameters, the single-variable method was used to investigate the influence of three important parameters: laser fluence, pulse frequency, and scanning speed on the cleaning efficacy.

Laser fluence ($F$) is an important parameter affecting cleaning quality, which can be coupled with laser power, spot diameter, and pulse frequency parameters. The laser fluence ($F$) density can be expressed by the Formula (1) [15]:

$$F = \frac{4P_v}{\pi f d^2} \tag{1}$$

where $P_v$ is the average laser power, $f$ is the pulse frequency, and $d$ is the diameter of the focused spot. The pulse frequency used in this experiment is 10 kHz and the diameter of the focused spot is 1 mm. Figure 5 shows the surface morphology of the sample after laser cleaning at different laser fluences. As shown in Figure 5a, there is a large amount of graphite residue on the surface of the sample with a laser fluence of 1.27 J/cm$^2$. The majority of the sample surface is still obscured by the black graphite lubrication coating, with only certain areas (indicated by the circled regions in the figure) revealing the magnesium alloy morphology. As the laser fluence escalates from 2.55 to 5.09 J/cm$^2$, the cleaning efficacy of the coating markedly improves. As shown in Figure 5b–d, the residual amount of graphite was reduced from the local blocky residue in Figure 5b to no obvious residue observable by the naked eye in Figures 5c and 6d, and the magnesium alloy substrate was exposed on the surface of the samples after cleaning. However, partial melting traces are observed in some scratch areas in Figure 5d. Figure 5e,f depict the cleaning outcomes at laser fluences of 6.37 J/cm$^2$ and 7.64 J/cm$^2$, respectively. At these higher fluences levels, while the graphite is entirely eliminated, excessive laser energy adversely impacts the substrate. In Figure 5e, melt marks are evident across much of the sample surface, with localized areas exhibiting a yellowish burn. Meanwhile, in Figure 5f, the entire matrix is covered by melt marks, accompanied by regular parallel stripe-like laser overlap marks This phenomenon indicates that the surplus laser energy has caused synchronous surface treatment on the exposed magnesium alloy substrate surface after removing the graphite coating within the laser range.

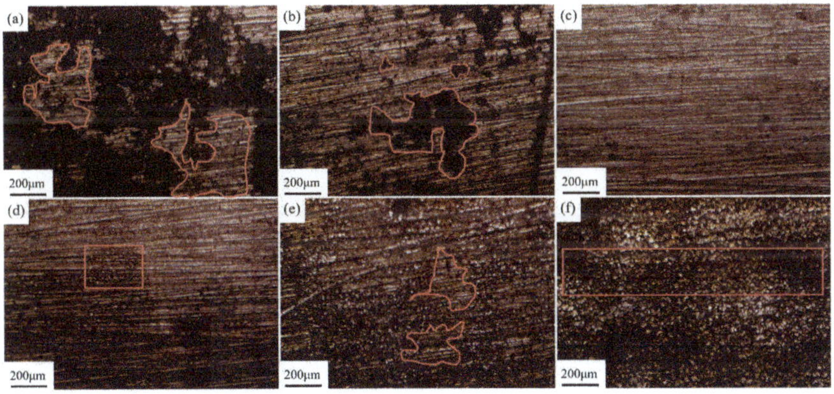


**Figure 5.** Macromorphology of sample surface after laser cleaning at different laser fluences: (**a**) 1.27 J/cm$^2$; (**b**) 2.55 J/cm$^2$; (**c**) 3.82 J/cm$^2$; (**d**) 5.09 J/cm$^2$; (**e**) 6.37 J/cm$^2$; (**f**) 7.64 J/cm$^2$.

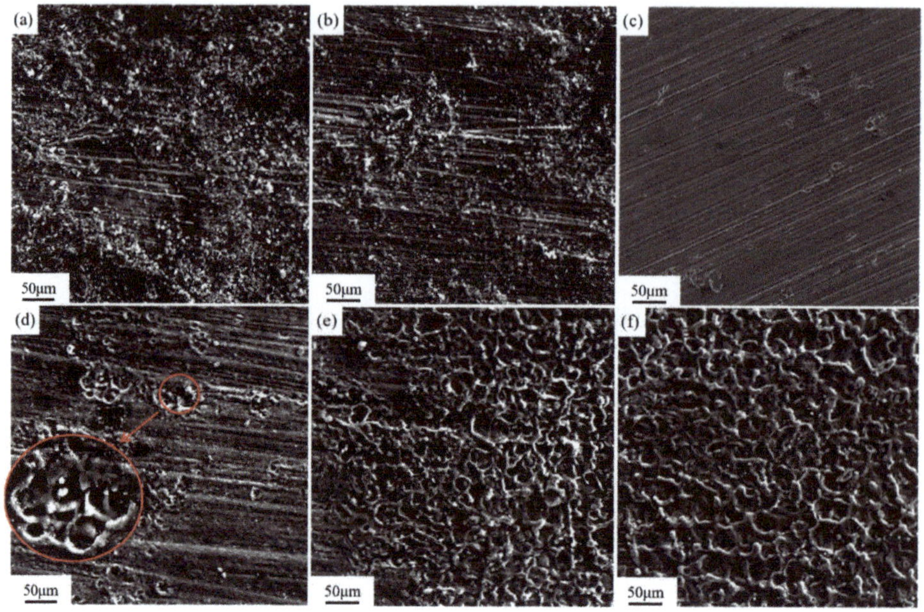


**Figure 6.** Microscopic morphology of sample surface after laser cleaning at different laser fluences: (**a**) 1.27 J/cm$^2$; (**b**) 2.55 J/cm$^2$; (**c**) 3.82 J/cm$^2$; (**d**) 5.09 J/cm$^2$; (**e**) 6.37 J/cm$^2$; (**f**) 7.64 J/cm$^2$.

Figure 6 shows the microstructure of the magnesium alloy surface after cleaning with different laser fluences observed under a scanning electron microscope. Compared with Figure 6a–f, as the laser fluence increases, the cleaning efficacy of the coating gradually improves. As shown in Figure 6c, the graphite is completely removed at 3.82 J/cm$^2$, revealing the morphology of the substrate, and the substrate is not affected by the laser. At 5.09 J/cm$^2$, although the graphite was completely removed, the magnesium alloy substrate began to appear slight melting marks at the scratch, indicating that the excess laser energy was also absorbed by the substrate surface, while the laser removed the graphite coating. When the laser is radiated to the surface of the sample, the flatness of the sample's surface will exert an impact on the absorption of light. The reflectivity at smooth positions is relatively higher. Because at the areas where scratches concentrate, the laser will undergo multiple reflections and absorptions within the grooves of the scratches. The absorption efficacy of the laser at these locations is relatively higher. The additional absorbed energy causes some parts of the scratched areas to reach the melting point first, resulting in the appearance of melting marks. Compared to other areas, it first reached the melting point of the magnesium alloy and appeared melting phenomenon. Figure 6e,f shows that when the laser energy is sufficiently high, the substrate surface absorbs enough energy, resulting in large-scale melting. The substrate surface reaches the melting temperature and then rapidly cools, resulting in an obvious cloud-like cooling and melting morphology on the metal surface. This indicates that the substrate has been damaged by excess laser energy, and the laser fluence at this time is too high. The process of laser cleaning shown in Figure 3 also proves this point. As shown in Figure 3, when the laser energy is too high, the area of excessive cleaning also increases with the increase of the laser scanning path. Due to the sufficient laser energy, the surface coating has been removed after scanning path 1. After path 2 moves, the originally cleaned area is damaged due to the overlapping of two adjacent light spots in the Y direction, and the exposed substrate absorbs excess energy, resulting in obvious black marks in high-speed photography images.

Figure 7 shows the EDS scanning results of the cleaned sample surface at different average powers. The surface of the uncleaned sample is composed of C (~63.43 wt%)—the main component of the graphite coating, O (~25.42 wt%) and Mg (~9.0 wt%), Zn, Si, and other elements. This study defines clearance rate as the percentage of the difference in surface graphite content before and after cleaning to the initial surface graphite content. As the average power increases from 1.27 to 1.27 $J/cm^2$, the C content decreases, while the Mg content gradually increases. Within the range of 1.27~3.82 $J/cm^2$, the C content exhibits a significant downward trend, decreasing from 63.43% before cleaning to 6.42%, the clearance rate reaches 89.88% at 3.82 $J/cm^2$. These data indicate that increasing laser fluence within this range significantly improves the cleaning efficacy of the lubrication coating. The curve tends to stabilize within the range of 3.82 to 7.64 $J/cm^2$, and the clearance rate remains at around 90%. The cleaning efficacy is optimal at the laser fluence of 5.09 $J/cm^2$, achieving a clearance rate of 91.46%, slightly higher than 89.88% at 3.82 $J/cm^2$.

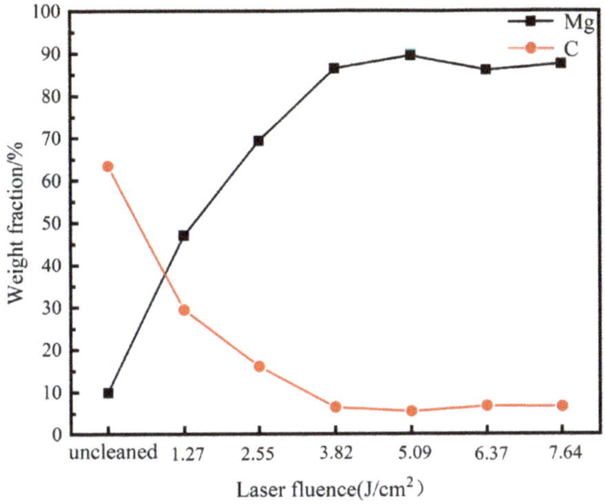


**Figure 7.** Surface element content of cleaned samples under different laser fluences.

Laser action can alter the microstructure of the substrate surface, which can impact the properties of the magnesium alloy substrate. Figure 8 displays the three-dimensional microstructure of the sample surface after cleaning with various laser fluences. Figure 9 illustrates the impact of laser fluence on the surface roughness of the cleaned sample. Surface roughness refers to the unevenness of a machined surface with relatively small intervals and tiny peaks and valleys. In this paper, the description of the surface roughness value refers to the arithmetic mean deviation Ra of the contour, and its measurement principle is the arithmetic mean of the absolute value of the contour deviation distance within the sampling length. As the laser fluence increases from 1.27 to 3.82 $J/cm^2$, the surface roughness of the sample decreases. However, as the laser fluence further increases from 3.82 to 7.64 $J/cm^2$, the surface roughness of the sample increases. The surface roughness of the 3.82 $J/cm^2$ cleaned sample is almost the same as that of the original magnesium alloy substrate surface, indicating that the cleaning efficacy is optimal at 3.82 $J/cm^2$. The graphite has been completely removed without causing any damage to the substrate. The surface roughness of the original substrate and the cleaned sample is significantly higher than that of the uncleaned sample. This phenomenon is due to a large number of scratches on the substrate caused by artificial sandpaper polishing, and the sprayed graphite particles adhere to the surface of the substrate to fill in the uneven scratches, making the surface flat. When the laser fluence is low, the laser cleaning efficacy is poor, and the remaining large graphite on the sample's surface increases the wave peak, resulting in higher surface roughness. As the laser fluence increases, the amount of residual graphite on the sample surface gradually decreases, leading to a reduction in roughness value. When the laser energy is too high, the excess energy is absorbed by the substrate, causing surface melting. Moreover, the higher the laser fluence, the more severe the melting becomes. Due to the short duration of pulsed laser action, the molten metal cools down before it can flow rapidly, leading to increased laser fluence and higher roughness.

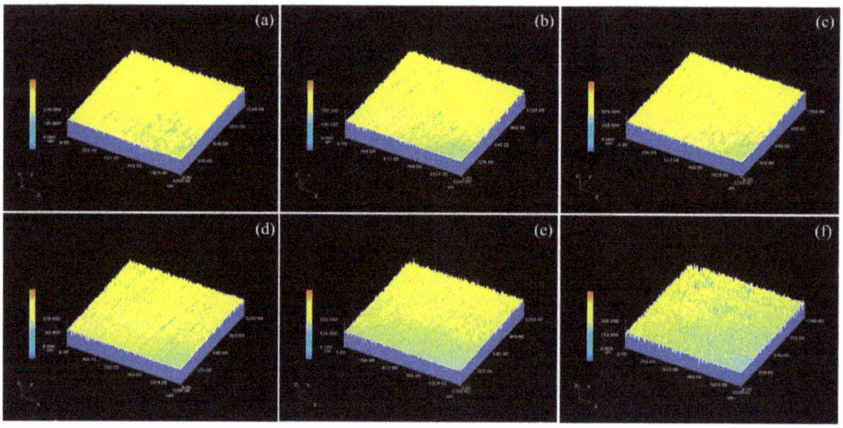


**Figure 8.** 3D Micromorphology of the sample surface after cleaning with different laser fluences: (**a**) 1.27 J/cm$^2$; (**b**) 2.55 J/cm$^2$; (**c**) 3.82 J/cm$^2$; (**d**) 5.09 J/cm$^2$; (**e**) 6.37 J/cm$^2$; (**f**) 7.64 J/cm$^2$.

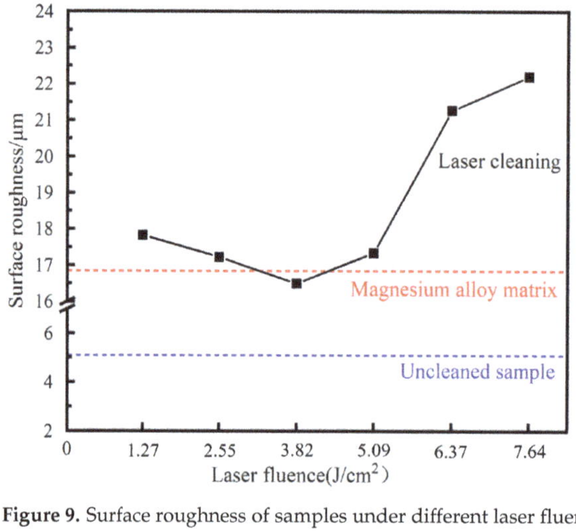


**Figure 9.** Surface roughness of samples under different laser fluences.

Based on the above analysis, as the laser fluence increases, the cleaning effect of the film coating gradually improves. However, it is not true that higher laser fluence always leads to better results. When the damage threshold of the substrate is exceeded, high-energy laser will cause damage to the substrate. Combining the cleaning effect and non-destructive considerations, the most suitable laser fluence for a 15 μm thick graphite lubricating coating on the surface of magnesium alloys is 3.82 J/cm$^2$.

### 3.3. The Effect of the Pulse Frequency on Cleaning

Pulse repetition frequency refers to the number of regular laser pulses output per unit time, which is the number of times pulses repeat within 1 second. When the laser power $P$ is constant, the single-pulse energy $P_S$ can be calculated based on the pulse frequency $f$, as shown in the Formula (2):

$$P_S = \frac{P}{f} \tag{2}$$

At a fixed laser power ($P$), the higher the pulse frequency ($f$), the smaller the output single-pulse energy ($P_S$).

Figure 10 shows the surface morphology of the samples after laser cleaning at different pulse frequencies. Three pulse frequencies of 10 kHz, 20 kHz, and 30 kHz were selected for laser cleaning experiments. In Figure 10a, it can be observed that after cleaning at 10 kHz, most of the bright magnesium alloy matrix is exposed on the surface of the sample, indicating a more effective cleaning outcome. In Figure 10b, after cleaning at 20 kHz, there is a significant increase in residual graphite, and the cleaned area shows no apparent metallic luster. As shown in Figure 10c, after cleaning at 30 kHz, the surface displays distinct rows of laser spot motion traces. The residual strip of graphite at the overlap of the two rows has not been removed, and the residual graphite coating in the area where the center energy of the laser spot moves higher is thicker than at 20 kHz.

**Figure 10.** Macromorphology of the sample surface after laser cleaning at different pulse frequencies: (**a**) 10 kHz; (**b**) 20 kHz; (**c**) 30 kHz. The area circled in red is strip shaped residual graphite traces after laser spot scanning.

Figure 11 shows the microstructure of the magnesium alloy surface after cleaning with different pulse frequencies observed under a scanning electron microscope. The information shown in Figure 11 is the same as that in Figure 10. As the pulse frequency increases, the cleaning efficacy of the membrane coating gradually decreases.

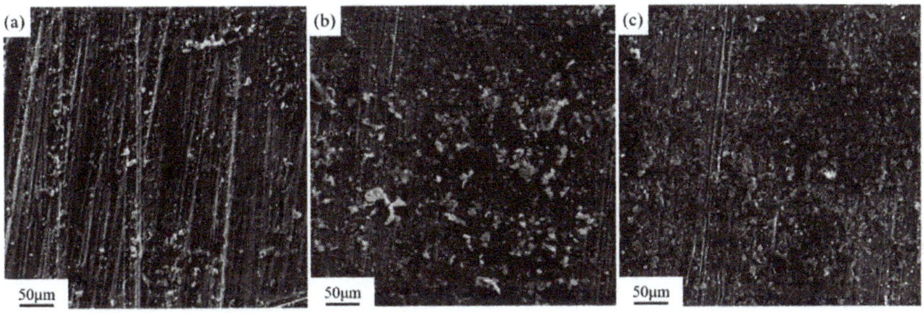


**Figure 11.** Microscopic morphology of sample surface after laser cleaning at different pulse frequencies: (**a**) 10 kHz; (**b**) 20 kHz; (**c**) 30 kHz.

Figure 12 shows the EDS scanning results of the cleaned sample surface at various pulse frequencies. As the pulse frequency increases from 10 to 30 kHz, the C content increases, while the Mg content decreases gradually. Under the pulse frequency of 10 kHz, the C content on the surface of the sample is 16.05%, and the clearance rate is 74.70%, which is 32.63% higher than that at 30 kHz. According to Figure 12, reducing the pulse frequency can significantly enhance the cleaning efficacy of the lubrication coating.

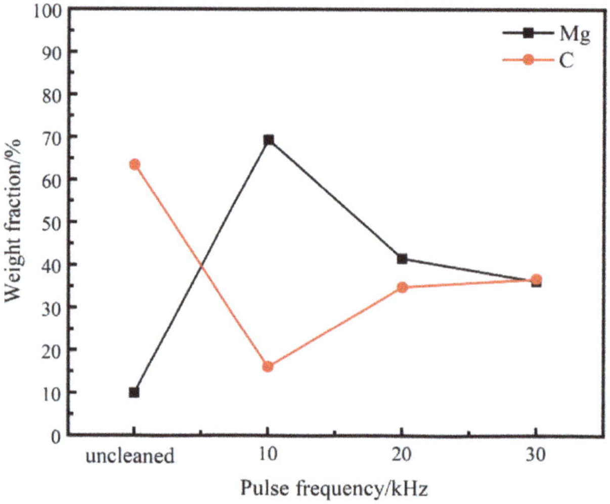


**Figure 12.** Surface element content of cleaned samples under different pulse frequencies.

Figure 13 shows the three-dimensional microstructure of the sample surface after cleaning with different pulse frequencies, while Figure 14 shows the effect of pulse frequency on the surface roughness of the cleaned sample. As the pulse frequency increases in the range of 10–30 kHz, the surface roughness of the sample increases and is higher than the original matrix, significantly higher than the surface of the uncleaned sample. When the laser power is constant, as the pulse frequency increases, the energy of a single laser pulse decreases, which is not enough to remove all lubricating coatings. The residual graphite on the surface of the sample leads to a decrease in surface smoothness and an increase in roughness.

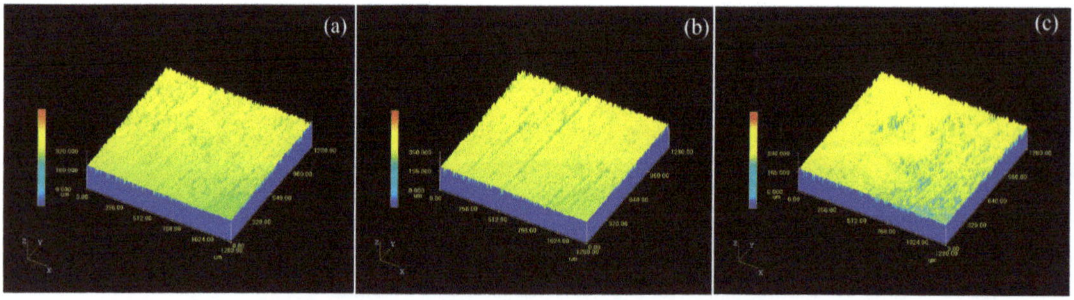


**Figure 13.** 3D Micromorphology of sample surface after cleaning with different pulse frequencies: (**a**) 10 kHz; (**b**) 20 kHz; (**c**) 30 kHz.

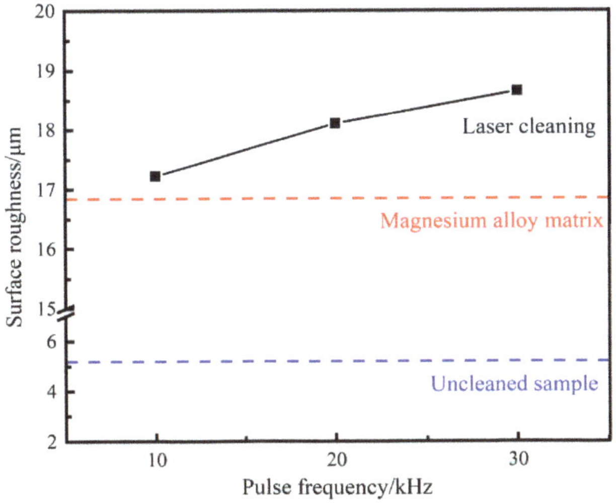


**Figure 14.** Surface roughness of samples under different pulse frequencies.

Based on the above analysis, as the pulse frequency increases, the cleaning efficacy of the membrane coating gradually decreases. Combining the cleaning effect and non-destructive considerations, the most suitable pulse frequencies for a 15 μm thick graphite lubricating coating on the surface of magnesium alloys is 10 kHz.

### *3.4. The Effect of the Scanning Speed on Cleaning*

Scanning speed refers to the speed at which the laser moves laterally along the X direction. The faster the scanning speed, the less time is required for scanning in the X direction, and the lower the overlap rate of the light spot in the X direction. The slower the scanning speed, the more time is required for scanning in the X direction, and the higher the overlap rate of the light spot in the X direction.

Figure 15 displays the macroscopic morphology of the sample surface after laser cleaning at various scanning speeds. Three scanning speeds of 1000 mm/s, 3000 mm/s, and 5000 mm/s were selected for cleaning experiments with a fixed laser fluence of 3.82 J/cm$^2$ and a pulse frequency of 10 kHz. Comparing Figure 15a–c, it can be seen that the cleaning efficacy is best at 3000 mm/s. The graphite coating on the surface of the sample can be completely removed, revealing the original substrate's bright morphology. At a speed of 1000 mm/s, clearly visible rows of laser motion marks appear on the surface, and the intersection line is burnt black. When the scanning speed is 5000 mm/s, a significant amount of residual graphite can be observed on the cleaned surface.

**Figure 15.** Macromorphology of sample surface after laser cleaning at different scanning speeds: (**a**) 1000 mm/s; (**b**) 3000 mm/s; (**c**) 5000 mm/s.

Figure 16 shows the microstructure of the sample surface after cleaning at different scanning speeds. In Figure 16a, a too-low speed (1000 mm/s) resulted in a high overlap rate of X-direction light spots, and the substrate surface absorbed too much residual energy, resulting in a large area of melting phenomenon. Obvious striped laser marks were seen in the microscopic morphology image. The best cleaning efficacy was achieved at a scanning speed of 3000 mm/s, where the graphite was completely removed and the substrate was not affected by the laser. At 5000 mm/s, the overlap rate of the X-direction spot was too low, and the laser energy acting on the unit area was small, with obvious traces of graphite residue visible on the surface. Based on the above analysis, it can be concluded that as the scanning speed decreases, the cleaning efficacy of the film coating gradually improves. However, too low a speed will lead to a high overlap rate of the laser spot in the X-direction, and the laser energy in the overlap area will be too high, damaging the substrate.

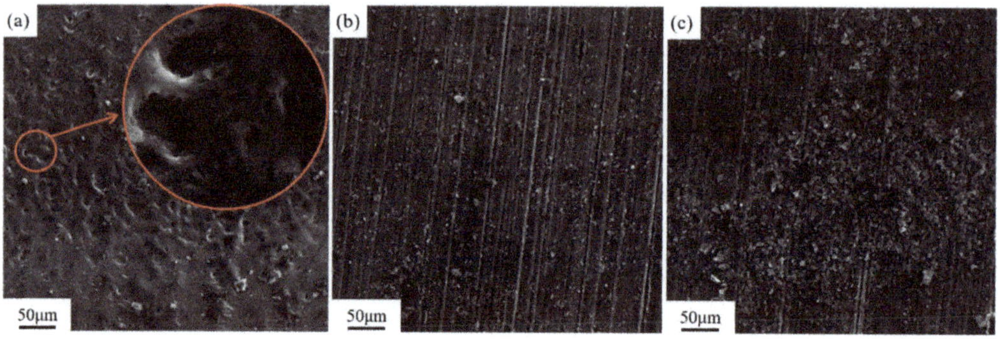


**Figure 16.** Microscopic morphology of sample surface after laser cleaning at different scanning speeds: (**a**) 1000 mm/s; (**b**) 3000 mm/s; (**c**) 5000 mm/s.

Figure 17 shows the EDS scanning results of the cleaned sample surface at different scanning speeds. The results indicate that as the scanning speed increases from 1000 to 5000 mm/s, the C content increases, while the Mg content decreases gradually. Under the scanning speed of 1000 mm/s, the content of C element decreased from 63.43% before cleaning to 5.77%, and the clearance rate reached 90.90%. The clearance rate was 89.88% at 3000 mm/s and 89.83% at 5000 mm/s. It can be seen that under the condition of an average power of 300 W, the scanning speed has little effect on the cleaning efficacy. As the scanning speed increases from 1000 to 5000 mm/s, the clearance rate decreases, but the slope is gentle and remains around 90%. Compared to the significant impact of average power and pulse frequency on the surface element content of the sample, the scanning speed is not considered to affect the cleaning efficacy of the lubricating coating significantly. Therefore, it is not the main focus of discussion.

Therefore, based on all the above analyses and considering the influence of laser parameters on surface morphology, element content, and roughness, as well as the cleaning effect and whether the substrate is damaged, it can be concluded that $F$ = 3.82 J/cm$^2$, $P$ = 300 W, $f$ = 10 kHz, and $v$ = 3000 mm/s are most suitable for cleaning the 15 μm thick graphite lubricating coating on the surface of the MB15 magnesium alloy.

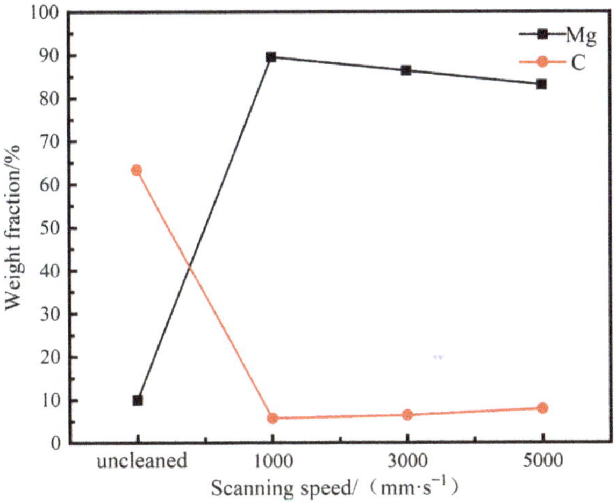


**Figure 17.** Surface element content of cleaned samples under different scanning speeds.

## 4. Conclusions

This study conducted laser cleaning experiments on the graphite lubrication coating on the surface of magnesium alloy using a nanosecond pulsed laser. The influence of laser power, pulse frequency, and scanning speed on the surface morphology, element content, and surface roughness of cleaned samples were studied. The innovation of this study was to verify the feasibility of using a nanosecond pulsed laser to clean the lubricating coating on the surface of forged specimens and elucidate the main mechanism of laser cleaning the coating. The main results are as follows:

(1) As the laser fluence increases, the energy density gradually increases, and the clearance rate improves. As the pulse frequency increases, the single pulse energy decreases and the clearance rate decreases. The scanning speed increases from 1000 to 5000 mm/s, and the spot overlap rate gradually decreases, while the clearance rate first increases and then decreases.

(2) Increasing laser power and reducing pulse frequency can significantly reduce C content and improve cleaning efficacy. The scanning speed (1000~5000 mm/s) has little effect on the changes in element content, and the C content has remained around 6%.

(3) Under the action of $F = 3.82\ \text{J/cm}^2, f = 10$ kHz, and $v = 3000$ mm/s, the graphite clearance rate reached 89.88% and the matrix was not damaged, which is beneficial for cleaning the 15 μm thick graphite lubricating coating of the MB15 magnesium alloy surface.

(4) The dominant mechanism for laser cleaning of graphite lubricating coatings on magnesium alloy surfaces is thermal ablation mechanism, and the higher the laser fluence, the more intense the ablation phenomenon and the better the cleaning effect.

**Author Contributions:** Conceptualization, Z.X., D.Z. and E.L.; Formal analysis, Y.Y., D.Z., S.X. and E.L.; Investigation, Y.Y.; Methodology, Y.Y. and S.X.; Project administration, J.X. and B.G.; Supervision, Z.X., D.S., J.X. and B.G.; Writing—original draft, Y.Y.; Writing—review and editing, Z.X. All authors have read and agreed to the published version of the manuscript.

**Funding:** This research was funded by CGN HIT Advanced Nuclear and New Energy Research Institute (Grant No. CGN-HIT202211), the National Key R&D Program of China (Program No. 2022YFB4601500), and the National Natural Science Foundation of China (Grant No. U2133214).

**Institutional Review Board Statement:** Not applicable.

**Informed Consent Statement:** Not applicable.

**Data Availability Statement:** The original contributions presented in this study are included in the article. Further inquiries can be directed to the corresponding author.

**Conflicts of Interest:** The authors declare no conflicts of interest.

## References


1. Park, K.B.; Choi, C.; Yu, H.; Chae, S.-R.; Kim, I.S. Optimization of chemical cleaning for reverse osmosis membranes with organ fouling using statistical design tools. *Environ. Eng. Res.* **2018**, *23*, 474–484. [CrossRef]
2. Seifert, S.; Kleiman, J.I.; Heimann, R.B. Thermal optical properties of plasmas prayed mullite coatings for space launch vehicl *Spacecr. Rocket.* **2006**, *43*, 439–442. [CrossRef]
3. Bonacchi, D.; Rizzi, G.; Bardi, U.; Scrivani, A. Chemical stripping of ceramic films of titanium aluminum nitride from hard me substrates, Surf. *Coatings Technol.* **2003**, *165*, 35–39. [CrossRef]
4. Ouyang, J.; Mativenga, P.; Goffin, N.; Liu, W.; Liu, Z.; Mirhosseini, N.; Jones, L.; Woolley, E.; Li, L. Energy consumption a performance optimisation of laser cleaning for coating removal. *CIRP. Manuf. Sci. Technol.* **2022**, *37*, 245–257. [CrossRef]
5. Kim, I.T.; Jung, Y.S.; Kim, K.J.; Lee, D.U. Effect of Blast Cleaning on Fatigue Behavior of Non-load-carrying Fillet Weld Cruciform Joints. *J. Korean Soc. Steel Constr.* **2009**, *21*, 55–62.
6. Bartha, L.; Koltay, E.; Mórik, G. Abrasive blast cleaning method for the renewal of worn-out acceleration tubes. *Nucl. Instru Methods Phys. Res. Sect. B Beam Interact. Mater. Atoms* **1996**, *111*, 157–160. [CrossRef]
7. Bergs, T.; Müller, U.; Vits, F.; Barth, S. Grinding wheel wear and material removal mechanisms during grinding of polycrystalli diamond. *CIRP Conf. Manuf. Syst.* **2020**, *93*, 1520–1525. [CrossRef]
8. Verhaagen, B.; Zanderink, T.; Rivas, D.F. Ultrasonic cleaning of 3D printed objects and cleaning challenge devices. *Appl. Acou* **2016**, *103*, 172–181. [CrossRef]
9. Zhao, W.; Yu, Z. Self-cleaning effect in high quality percussion ablating of cooling hole by picosecond ultra-short pulse laser. *O Lasers Eng.* **2018**, *105*, 125–131. [CrossRef]
10. Genna, S.; Lambiase, F.; Leone, C. Effect of laser cleaning in laser assisted joining of CFRP and PC sheets. *Compos. Part B E* **2018**, *145*, 206–214. [CrossRef]
11. Rauh, B.; Kreling, S.; Kolb, M.; Geistbeck, M.; Boujenfa, S.; Suess, M.; Dilger, K. UV-laser cleaning and surface characterization an aerospace carbon fibre reinforced polymer. *Int. J. Adhes. Adhes.* **2018**, *82*, 50–59. [CrossRef]
12. Jia, Y.H.; Gong, X.; Peng, P.; Wang, Z.; Tian, Z.; Ren, L.; Fu, Y.; Zhang, H. Toward high carrier mobility and low contact resistan laser cleaning of PMMA residues on graphene surfaces. *Nano-Micro Lett.* **2016**, *8*, 336–346. [CrossRef] [PubMed]
13. Hu, Y.; Liu, C.; Li, K.; Cheng, J.; Zhang, Z.; Han, E. An efficient laser decontamination process based on non-radioactive specime of nuclear power materials. *Materials* **2023**, *16*, 7643. [CrossRef] [PubMed]
14. Ready, J. Effects Due to Absorption of Laser Radiation. *J. Appl. Phys.* **1965**, *36*, 462–468. [CrossRef]
15. Li, Z.; Zhang, D.; Su, X.; Yang, S.; Xu, J.; Ma, R.; Shan, D.; Guo, B. Removal mechanism of surface cleaning on TA15 titanium all using nanosecond pulsed laser. *Opt. Laser Technol.* **2021**, *139*, 106998. [CrossRef]
16. Tsunemi, A.; Endo, A.; Ichishima, D. Paint removal from aluminum and composite substrate of aircraft by laser ablation usi TEA $CO_2$ lasers. *High-Power Laser Ablation Int. Soc. Opt. Photonics* **1998**, *3343*, 1018–1022.
17. Kumar, A.; Sapp, M.; Vincelli, J.; Gupta, M.C. A study on laser cleaning and pulsed gas tungsten arc welding of Ti–3Al–2.5 alloy tubes. *J. Mater. Process. Technol.* **2010**, *210*, 64–71. [CrossRef]
18. Liu, K.; Garmire, E. Paint removal using lasers. *Appl. Opt.* **1995**, *34*, 4409–4415. [CrossRef]
19. Rodríguez, A.; Lopez, A.J.; Lamas, J.; Moreno, A.; Ramil, A. Robot-assisted laser ablation for 3D surfaces. Application for pai removal with ultrashort pulse laser. *Opt. Lasers Eng.* **2023**, *160*, 107284. [CrossRef]
20. Li, J.; Yang, J.; Liu, J.; Chen, H.; Duan, Y.; Pan, X. Research on laser cleaning technology for aircraft skin surface paint lay *Materials* **2024**, *17*, 2414. [CrossRef]
21. Chen, G.X.; Kwee, T.J.; Tan, K.P.; Choo, Y.S.; Hong, M.H. High-power fibre laser cleaning for green shipbuilding. *J. Las Micro/Nanoeng.* **2012**, *7*, 249–253. [CrossRef]
22. Kholodova, S.I.; Goryachkin, D.A.; Koval'chuk, L.V. The cleaning of works of art made from metal with laser radiation wavelengths of 1.06 and 10.6μm. *J. Opt. Technol.* **2010**, *77*, 303–308. [CrossRef]
23. Ye, Y.; Yuan, X.; Xiang, X.; Cheng, X.; Miao, X. Laser cleaning of particle and grease contaminations on the surface of optic *Opt.-Int. J. Light Electron Opt.* **2012**, *123*, 1056–1060. [CrossRef]
24. Harris, C.D.; Shen, N.; Rubenchik, A.M.; Demos, S.G.; Matthews, M.J. Characterization of laser-induced plasmas associated wi energetic laser cleaning of metal particles on fused silica surfaces. *Opt. Lett.* **2015**, *40*, 5212–5215. [CrossRef] [PubMed]
25. Yue, L.; Wang, Z.; Guo, W.; Li, L. Axial laser beam cleaning of tiny particles on narrow slot sidewalls. *J. Phys. D Appl. Phys.* **20** *45*, 365106–365113. [CrossRef]

Chloros, J.; Salmon, H.; Talland, V. Laser cleaning at the Isabella Stewart Gardner Museum, Boston, USA, Sixteen Roman sculptures, fourteen months, and three conservators. *Stud. Conserv.* **2015**, *60* (Suppl. 1), S41–S48. [CrossRef]
Li, Y.; Ye, Y.; Liu, H.R.; Shen, R.; Yuan, X.; Miao, X.; Yao, C.; Bai, Q.; Lu, L.; Zhang, P. Time resolved imaging for investigating laser-material interactions during laser irradiation cleaning on murals. *Opt. Laser Technol.* **2023**, *157*, 108679. [CrossRef]
Pozo-Antonio, J.S.; Papanikolaou, A.; Philippidis, A.; Melessanaki, K.; Rivas, T.; Pouli, P. Cleaning of gypsum-rich black crusts on granite using a dual wavelength Q-Switched Nd, YAG laser. *Constr. Build. Mater.* **2019**, *226*, 721–733. [CrossRef]
Zhang, G.; Hua, X.; Huang, Y.; Zhang, Y.; Li, F.; Shen, C.; Cheng, J. Investigation on mechanism of oxide removal and plasma behavior during laser cleaning on aluminum alloy. *Appl. Surf. Sci.* **2019**, *506*, 144666. [CrossRef]
Zhang, D.; Xu, J.; Li, Z.; Li, K.; Wang, C.; Shan, D.; Guo, B. Removal mechanism of blue paint on aluminum alloy substrate during surface cleaning using nanosecond pulsed laser. *Opt. Laser Technol.* **2022**, *149*, 107882. [CrossRef]
Wang, Z.; Zeng, X.; Huang, W. Parameters and surface performance of laser removal of rust layer on A3 stee. *Surf. Coat. Technol.* **2003**, *166*, 10–16. [CrossRef]
Ye, Y.; Yuan, X.; Xiang, X.; Dai, W.; Chen, M.; Miao, X.; Lv, H.; Wang, H.; Zheng, W. Laser plasma shockwave cleaning of SiO2 particles on gold film. *Opt. Lasers Eng.* **2011**, *49*, 536–541. [CrossRef]
Asmus, J.F. Light for art conservation. *Interdiscip. Sci. Rev.* **1987**, *12*, 171. [CrossRef]
Rana, M.T.; Hussein, A.J. Numerical modeling of 193-nm excimer laser ablation on CR-39 polymer. *Electron. Mater.* **2019**, *48*, 977–981.
Watkins, K.G.; Chen, X.; Fujioka, T.; Matsunawa, A. Mechanisms of laser cleaning. *Proc. SPIE-Int. Soc. Opt. Eng.* **2000**, *3888*, 165–174.
Psyllaki, P.; Oltra, R. Preliminary study on the laser cleaning of stainless steels after high temperature oxidation. *Mater. Sci. Eng. A* **2000**, *282*, 145–152. [CrossRef]

*materials*

MDPI

*Communication*

# Effects of Process Parameters on Forming Quality and Microstructure of FeCrAl-ODS Alloy Fabricated by Selective Laser Melting

**Shenghua Zhang [1,*], Fudong Li [1], Yu Wang [1], Hongwen Su [2] and Jun Li [3,*]**

[1] School of Materials Science and Engineering, North University of China, Taiyuan 030051, China; 18298897601@163.com (F.L.); wangyu@nuc.edu.cn (Y.W.)
[2] Shanxi Huachuang Laser Technology Co., Ltd., Changzhi 046000, China; sxszshw@163.com
[3] Taiyuan Iron and Steel (Group) Co., Ltd., Taiyuan 030051, China
* Correspondence: shenghuazhang@nuc.edu.cn (S.Z.); lijun2@tisco.com.cn (J.L.)

**Abstract:** This study systematically investigated the effects of selective laser melting (SLM) process parameters on the forming quality and microstructure of FeCrAl oxide dispersion-strengthened (ODS) alloy. Through orthogonal experimental design, the influences of laser power (300–320 W), scanning speed (650–850 mm/s), and hatch spacing (0.05–0.07 mm) on the surface morphology and internal defects of as-built samples were analyzed. The microstructural evolution under different volumetric energy densities (VED) was also analyzed. The results indicate that hatch spacing significantly affected crack and pore formation, with minimal defects observed at 0.06 mm. Excessive laser power (320 W) or VED (318.0 J/mm$^3$) led to elevated melt pool temperatures, causing element evaporation, grain coarsening, and <100> preferential oriented texture, thereby reducing hardness to 234 HV. The optimal parameters—laser power of 310 W, scanning speed of 650 mm/s, and hatch spacing of 0.06 mm (VED 265.0 J/mm$^3$)—yielded the highest hardness (293 HV), fine-grained structures, and a high proportion of low-angle grain boundaries (LAGBs) with significant residual stress. This research provides a theoretical foundation for optimizing SLM processes for FeCrAl-ODS alloys.

**Keywords:** FeCrAl-ODS alloy; selective laser melting; process parameter; forming quality; microstructure

## 1. Introduction

FeCrAl-ODS alloys exhibit exceptional performance in high-temperature environments, characterized by superior oxidation resistance, robust tolerance to irradiation, and enhanced mechanical properties, making them promising candidates for nuclear reactor cladding and high-temperature structural components [1,2]. However, traditional powder metallurgy methods for fabricating ODS alloys face limitations, including complex processing, high costs, and challenges in forming complex geometries [3,4]. Additionally, subsequent rolling processes often fail to eliminate elongated grain structures, leading to anisotropic properties [5]. To address these issues, emerging techniques such as additive manufacturing (AM) have been explored [6–10]. Among these, selective laser melting (SLM), an AM technology, enables the direct fabrication of high-precision complex parts through layer-by-layer melting. Its rapid cooling rates and micro-melt pool characteristics facilitate the uniform dispersion of nano-sized oxide particles and inhibit oxide particles' coarsening [11]. Nevertheless, the influence of SLM process parameters on the

 https://doi.org/10.3390/ma1811246

microstructure and properties of FeCrAl-ODS alloys with high oxide content remains insufficiently studied.

Prior studies highlight volumetric energy density (VED) as a critical factor in controlling melt pool dynamics, defect formation, and grain structure during SLM [12,13]. For instance, Ahmed et al. [14] found that, when processing the crack-susceptible and unweldable high-temperature nickel-based alloy CM247LC, laser power and hatch spacing had minimal effects on crack formation. However, optimizing hatch spacing contributed to a significant reduction in cracking. Liu et al. [15] reported that, during the SLM processing of SUS316L, scanning speed had a major influence on porosity. Li et al. [16] observed that a high scanning speed or low laser power promoted the transition from columnar to equiaxed grains. Greco et al. [17] studied the effect of laser power under constant VED conditions, and found that increasing the laser power reduced surface roughness perpendicular to the build direction, while simultaneously increasing relative density and microhardness. Yang et al. [18] reported that excessive VED increases porosity and cracks in 316 L stainless steel, while Zhang et al. [19] demonstrated that nano-oxide particles suppress phase transformation via a pinning effect. However, research on the SLM process optimization and microstructural evolution of FeCrAl-ODS alloys, particularly those with high oxide content, is still limited. This study systematically investigates the effects of laser power, scanning speed, and hatch spacing on the forming quality and microstructure of an FeCrAl-ODS alloy containing 0.91 wt. % $Y_2O_3$. The higher addition of $Y_2O_3$ is intended to improve the mechanical property of FeCrAl-ODS in this study. Through metallographic analysis, electron backscatter diffraction (EBSD), and microhardness testing, the correlation between process parameters and mechanical performance is elucidated, providing theoretical guidance for the additive manufacturing of high-performance FeCrAl-ODS alloys.

## 2. Experimental Methods

FeCrAl alloy powder with a particle size range of 15–53 μm and $Y_2O_3$ powder with a particle size of 30 nm were used in the experiment. The FeCrAl-ODS alloy powder with 1 wt. % yttria addition was prepared via low-energy ball milling. Stainless steel balls with diameters of 10 mm and 5 mm, blended in a 1:1 weight ratio, are used for ball milling. The milling parameters were as follows: a ball-to-powder ratio of 5:1, a milling duration of 4 h, and a rotational speed of 100 rotation per minute. To prevent powder contamination, the mixed powder was ball-milled under high-vacuum conditions (<0.1 Pa). Contamination originating from the stainless pot and balls is negligible under low-energy ball milling conditions. The morphology and surface elemental distribution of the as-milled powders were analyzed via a Zeiss Sigma 300 field-emission scanning electron microscope (SEM, Carl Zeiss AG, Oberkochen, Germany) integrated with energy-dispersive X-ray spectroscopy (EDS), as shown in Figure 1. The FeCrAl alloy powders maintained good sphericity after milling. A magnified view of a single powder particle within the red box in Figure 1 is shown in the inset, revealing that the nano-sized $Y_2O_3$ particles adhered to the surface of the FeCrAl alloy powders after milling. Elemental mappings indicate that the $Y_2O_3$ particles were relatively uniformly distributed on the powder surface. A small fraction of irregularly shaped particles was observed; however, the overall powder morphology meets the spreading requirements for SLM processing. The size of oxide particles in the as-built samples was 30–40 nm, and they distributed uniformly in the matrix, according to the prior study [11].

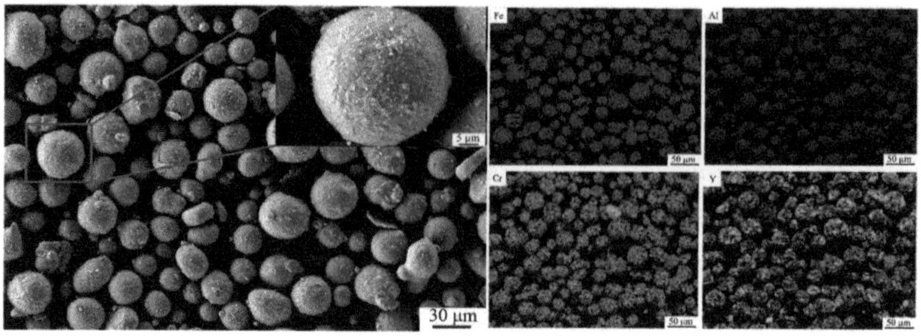


**Figure 1.** Second electron images and elemental mappings of FeCrAl-ODS alloy powder.

The cubic specimens (10 × 10 × 10 $mm^3$) were fabricated via a commercial SLM system (EP-M150, E plus 3D Technology Co., Ltd, Beijing, China) equipped with a fiber laser (spot diameter: 70 μm) capable of achieving a maximum scanning speed of 8 m/s. To minimize anisotropic defects, a 67° interlayer rotation strategy was applied during scanning. The thickness L of the powder spread was 30 μm. The substrate made of 304 stainless steel was preheated to 120 °C prior to processing, and argon shielding gas was utilized to maintain an oxygen concentration below 0.1 vol.% within the build chamber throughout the forming stage. Based on previous studies [11], the printing parameter ranges were selected as follows: laser power (*P*) of 300–320 W, scanning speed (*v*) of 650–850 mm/s, and hatch spacing (*H*) of 0.05–0.07 mm. The Y content in the fabricated alloy samples was determined to be 0.70 wt. % using inductively coupled plasma optical emission spectroscopy (ICP-OES, Agilent 720ES, Agilent Technologies Inc., Santa Clara, CA, USA), corresponding to $Y_2O_3$ contents of 0.91 wt. %. The analysis of Y content was performed with the following parameters: RF power of 1.2 KW, plasma flow of 15 L/min, auxiliary flow of 1.5 L/min, nebulizer flow of 0.75 L/min, sample uptake delay of 15 s, and 3 replicates. An inverted metallurgical microscope (MDS 400) was employed to observe the morphology and distribution of defects in the mechanically ground and polished samples. Electrolytic polishing was conducted using a solution of 10% $HClO_3$ and 90% $C_2H_5OH$, and electron backscatter diffraction (EBSD) was used to characterize the grain size and crystallographic texture of the specimens fabricated under different VED. The EBSD data were processed using Channel 5 software. The microhardness of each sample was measured 5 times using an HV-1000B Vickers (Laizhou Hengyi Testing Instrument Co., Ltd., Laizhou city, China) hardness tester under a load of 0.2 kgf, with a dwell time of 15 s.

## 3. Results and Discussion

### 3.1. *Effect of Process Parameters on the Forming Quality*

A three-factor, three-level orthogonal experimental design was employed, as shown in Table 1. The VED was calculated using the formula VED = $P/vHL$. Top-view images of the fabricated samples are presented in Figure 2. Under the selected processing parameters, all specimens were successfully fabricated without any observable cracking on the surface. Notably, sample 8 exhibited a relatively rough surface with visible pits, as indicated by the red circles in Figure 2. This may be attributed to the high volumetric energy density resulting from the combination of high laser power and small hatch spacing during the scanning processes. Excessive energy input can lead to elevated melt pool temperatures, which may cause the evaporation of alloying elements or spattering of the powder [13]. As layers accumulate, such effects can result in uneven surface morphology. Interestingly, compared to sample 8, sample 1 had the same hatch spacing and a slower scanning speed, resulting in a higher corresponding volumetric energy density, yet no surface pits were observed on its top surface. This suggests that the high energy density caused by reduced scanning speed has a relatively minor impact on forming quality, whereas energy density increases due to elevated laser power have a more significant effect on surface integrity.

**Table 1.** SLM process parameters of orthogonal experimental design.

| Sample ID | *P* (W) | *v* (mm/s) | *H* (mm) | VED (J/mm$^3$) |
|---|---|---|---|---|
| 1 | 300 | 650 | 0.05 | 307.7 |
| 2 | 300 | 750 | 0.06 | 222.2 |
| 3 | 300 | 850 | 0.07 | 168.1 |
| 4 | 310 | 650 | 0.06 | 265.0 |
| 5 | 310 | 750 | 0.07 | 196.8 |
| 6 | 310 | 850 | 0.05 | 243.1 |
| 7 | 320 | 650 | 0.07 | 234.4 |
| 8 | 320 | 750 | 0.05 | 284.4 |
| 9 | 320 | 850 | 0.06 | 209.2 |

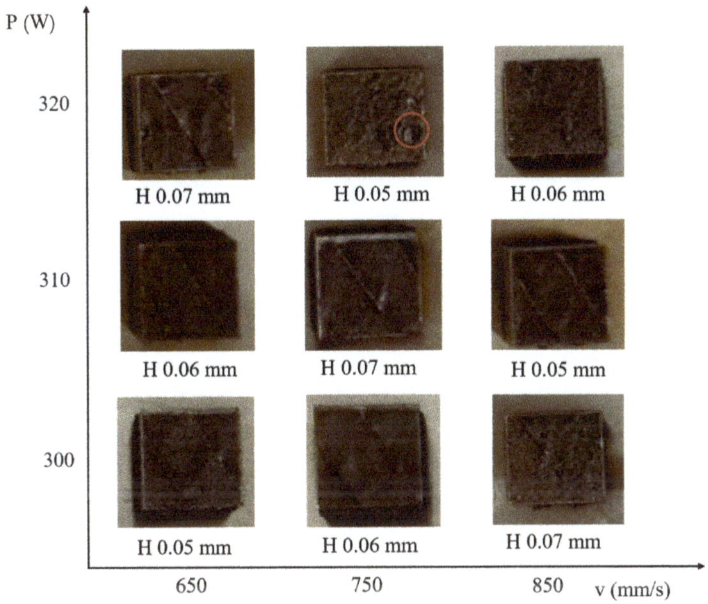


**Figure 2.** Photos of as-built samples.

Metallographic images provide a more intuitive means of observing the types and distributions of internal defects in the as-built blocks. Figure 3 presents the metallographic micrographs of the nine printed samples. It can be observed that when the hatch spacing is 0.07 mm, a significant number of pores and extensive cracking are present. In contrast, when the hatch spacing is reduced to 0.06 mm, the number of cracks is significantly reduced. However, when the hatch spacing further decreases to 0.05 mm, cracks appear. This indicates that crack formation is closely related to hatch spacing. A larger hatch spacing results in a lower energy density, leading to poor overlap between adjacent melt tracks and the formation of more cracks. When the hatch spacing is too low, hot cracking susceptibility increases. Additionally, unmelted powder can also be observed, as shown in Figure 3a. The porosity percentages are quantitatively shown in red font at the bottom left of each figure. When the hatch spacing was 0.05 mm, more pores (>1.7%) were observed within the samples, particularly at a laser power of 320 W, where both the number and size of pores were the greatest, as shown in Figure 3b. These large, round pores are typically formed by the evaporation of light elements, indicating that a laser power of 320 W is excessive for a hatch spacing of 0.05 mm. When the hatch spacing was increased to 0.06 mm, the number of defects in the fabricated parts was significantly reduced. At a laser power of 310 W and a scanning speed of 650 mm/s, only a small number of fine pores (0.5%) were observed in the sample. Therefore, for the alloy system studied in this work, the optimal processing parameters are a laser power of 310 W, scanning speed of 650 mm/s, and hatch spacing of 0.06 mm.

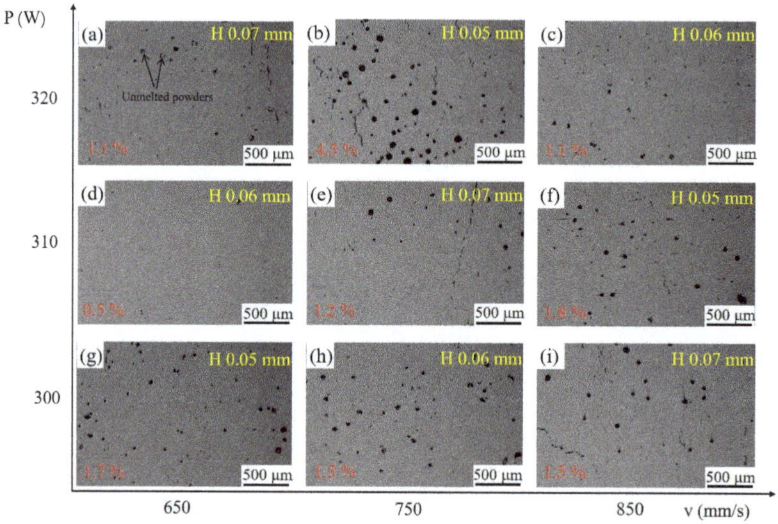


**Figure 3.** Metallographic micrographs of as-built samples. (**a**) P (320 W), v (650 mm/s), H (0.07 mm); (**b**) P (320 W), v (750 mm/s), H (0.05 mm), (**c**) P (320 W), v (850 mm/s), H (0.06 mm); (**d**) P (310 W), v (650 mm/s), H (0.06 mm), (**e**) P (310 W), v (750 mm/s), H (0.07 mm); (**f**) P (310 W), v (850 mm/s), H (0.05 mm), (**g**) P (300 W), v (650 mm/s), H (0.05 mm); (**h**) P (300 W), v (750 mm/s), H (0.06 mm), (**i**) P (300 W), v (850 mm/s), H (0.07 mm).

### *3.2. Effect of Process Parameters on the Evolution of Microstructure*

In order to obtain a clear evolution of the microstructure with VED variation, three samples with around 60 J/mm$^3$ increments were designed. The *P* was set as constant to reduce the complications associated with different process parameters. To investigate the effect of VED on the microstructure of SLM-fabricated alloys, a low VED of 202.6 J/mm$^3$ and a high VED of 318.0 J/mm$^3$ were selected based on the optimal processing condition, which corresponded to a VED of 265.0 J/mm$^3$. The specific processing parameters are listed in Table 2. Metallographic images of the cross-sections of the etched samples are shown in Figure 4. After etching, distinct melt pool tracks could be observed in all three samples. As the VED increased due to a reduction in scanning speed, the width of the melt pools significantly increased, and columnar grains aligned parallel to the laser scanning direction appeared at the melt track boundaries, as shown in Figure 4a,b. In contrast, when VED increased due to a decrease in hatch spacing, the melt pool width became narrower, as observed in Figure 4b,c. This indicates that the melt pool width is influenced by both the heat input and the hatch spacing. Additionally, grain coarsening was observed at the melt pool boundaries with increasing VED.

**Table 2.** Specific SLM process parameters of samples with different VED.

| Sample ID | *P* (W) | *v* (mm/s) | *H* (mm) | VED (J/mm$^3$) |
|---|---|---|---|---|
| 1 | 310 | 850 | 0.06 | 202.6 |
| 2 | 310 | 650 | 0.06 | 265.0 |
| 3 | 310 | 650 | 0.05 | 318.0 |

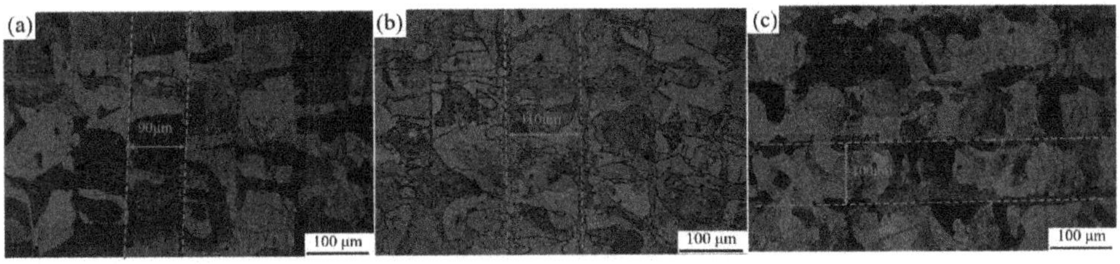


**Figure 4.** Metallographic micrographs of samples' cross-sections with different VEDs—(**a**) 202.6 J/mm$^3$, (**b**) 265.0 J/mm$^3$, (**c**) 318.0 J/mm$^3$.

Figure 5 presents the inverse pole figures (IPF), grain boundary (GB) maps, deformed–recrystallized (DefRex) maps, and kernel average misorientation (KAM) maps of samples printed under different VED conditions. With increasing VED, the proportion of red-colored grains in the IPF gradually increased, and the maximum texture intensity of the samples became more pronounced. This indicates an enhanced preferential orientation along the <100> direction with increasing heat input. This phenomenon is attributed to the elevated melt pool temperature and reduced cooling rate at higher VED levels, which allow a larger portion of the solidified section to reach recrystallization temperatures. As a result, a pronounced <100> preferential orientation formed in the cubic structured alloy. Consequently, excessively high VED may induce anisotropy in the mechanical properties of SLM-fabricated samples, potentially limiting their practical applications. Most of the grains in the three samples exhibit columnar morphology. When the VED exceeds 300 J/mm$^3$, the excessively high melt pool temperature leads to grain coarsening and the abnormal growth of some substructured grains.

In the GB maps, green lines represent LAGBs, while black lines indicate high-angle grain boundaries (HAGBs). As VED increased, the proportion of LAGBs initially rose and then decreases, a trend that is also reflected in the corresponding KAM maps, where the internal strain follows a similar pattern. This is because, in general, a higher proportion of LAGBs indicates a higher dislocation density and greater internal stress. Due to the extremely high cooling rate during the SLM process ($10^6$–$10^7$ K/s) [12], the solidified grains were predominantly deformed grains. At a VED of 202.6 J/mm$^3$, the melt pool temperature was relatively low, resulting in smaller grain sizes and moderate levels of LAGBs and internal strain. The heat-affected zone adjacent to the melt pool was narrow, and most of the microstructure consisted of deformed grains. When the VED increased to 265.0 J/mm$^3$, the melt pool temperature rose. However, due to the pinning effect of the nano-sized oxide particles, grain growth was effectively suppressed, and the transition from deformed grains to substructured grains was inhibited. This led to the accumulation of internal stress and an increased fraction of LAGBs, as the stored energy from rapid solidification was not released. At a higher VED of 318.0 J/mm$^3$, the proportion of HAGBs increased sharply, accompanied by a significant decrease in internal strain, as shown in the KAM maps. This may be attributed to the enhanced thermal driving force at elevated temperatures overcoming the pinning effect of the oxide particles, thereby promoting the transformation of a large number of deformed grains into substructured grains.

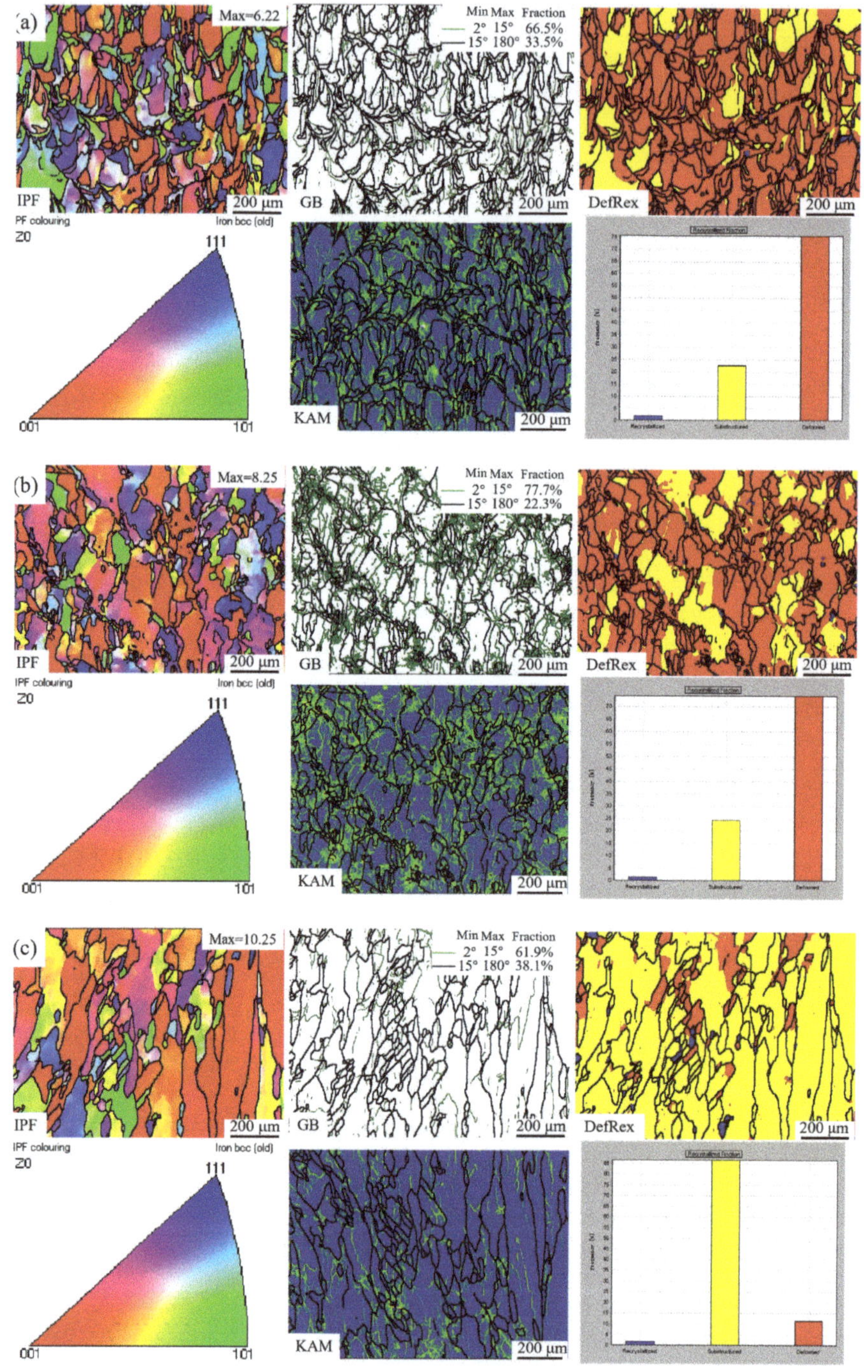


**Figure 5.** IPF, GB maps, DefRex maps, IPF coloring, KAM maps, and the recrystallized fraction of samples with different VEDs—(**a**) 202.6 J/mm$^3$, (**b**) 265.0 J/mm$^3$, (**c**) 318.0 J/mm$^3$.

### 3.3. *Effect of VED on the Hardness of FeCrAl-ODS Alloys*

During the SLM process, the VED directly determines the cooling rate, melt pool size, and thermal cycling frequency of the alloy material, thereby influencing the evolution of the microstructure and ultimately leading to differences in mechanical properties. The hardness of the samples fabricated under three different VED conditions was measured, and the comparison is shown in Figure 6. Owing to its finer grain size, high dislocation density, and significant internal stress, the sample fabricated at a VED of 265.0 J/mm$^3$ exhibited the highest hardness, reaching 293 HV. At a lower VED of 202.6 J/mm$^3$, the reduced internal stress and dislocation density resulted in a decreased hardness of approximately 263 HV. In contrast, at a high VED of 318.0 J/mm$^3$, grain coarsening and the release of internal stress led to the lowest hardness value, around 230 HV. Therefore, the hardness of SLM-fabricated FeCrAl-ODS alloys is closely correlated with the microstructural evolution induced by the scanning parameters. An appropriate VED can optimize the alloy hardness by controlling grain size, the fraction of substructured grains, and internal stress. The hardness values of FeCrAl-ODS alloys with 0.1 wt. % $Y_2O_3$ addition fabricated by SLM and conventional FrCrAl-ODS were 275 HV and around 300 HV [11,20]. This means that the hardness increases with an increase in $Y_2O_3$ addition for printed FeCrAl-ODS alloys, and their mechanical properties are comparable with those of the FeCrAl-ODS alloys fabricated via powder metallurgy methods. It is expected that a proper post-process heat treatment could further achieve a balance between strength and ductility in the alloy.

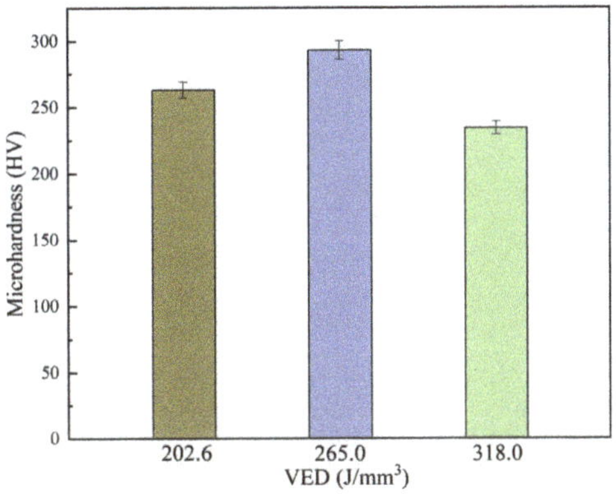


**Figure 6.** Microhardness comparison of samples with different VEDs.

## 4. Conclusions

By optimizing SLM process parameters, FeCrAl-ODS alloys with high forming quality and superior hardness were successfully fabricated. The evolution of the microstructure under varying VED values was investigated, and the following conclusions were drawn:

(1) Optimization of forming quality—Hatch spacing critically influenced defect distribution. At 0.06 mm, overlapping melt tracks minimized cracks and pores, whereas excessive laser power (320 W) or VED (318.0 J/mm$^3$) caused surface pits due to element evaporation from overheating. The scanning speed has a relatively minor impact on forming quality;

(2) Microstructural regulation—VED significantly affected grain structure and texture. At 265.0 J/mm$^3$, nano-sized $Y_2O_3$ particles inhibited grain coarsening via pinning effects, forming fine grains with a high fraction of LAGBs and achieving a peak hardness of 293 HV. Conversely, excessive VED (318.0 J/mm$^3$) induced abnormal

grain growth and released internal stress, reducing hardness to 230 HV. A moderate VED is conducive to control microstructure evolution;

(3) Optimal parameters—The combination of 310 W laser power, 650 mm/s scanning speed, and 0.06 mm hatch spacing (VED 265.0 $J/mm^3$) balanced forming quality and mechanical performance, laying a foundation for engineering applications of FeCrAl-ODS alloys.

Future work should focus on optimizing strength–ductility trade-offs through post-processing heat treatments and exploring residual stress control mechanisms during complex component fabrication.

**Author Contributions:** Conceptualization, S.Z.; methodology, F.L.; software, F.L.; formal analysis, S.Z. and F.L.; investigation, F.L., Y.W. and H.S.; writing—original draft preparation, S.Z. and F.L.; writing—review and editing, Y.W. and J.L.; supervision, S.Z. and J.L. All authors have read and agreed to the published version of the manuscript.

**Funding:** This research was funded by the Shanxi Scholarship Council of China (No. 2023-142), the Shanxi Province Key Research and Development Project (No. 202302050201012), and the special fund for Science and Technology Innovation Teams of Shanxi Province (No. 202304051001015).

**Institutional Review Board Statement:** Not applicable.

**Informed Consent Statement:** Not applicable.

**Data Availability Statement:** The original contributions presented in this study are included in the article. Further inquiries can be directed to the corresponding authors.

**Conflicts of Interest:** Author Hongwen Su was employed by the company Shanxi Huachuang Laser Technology Co., Ltd. Author Jun Li was employed by the company Taiyuan Iron and Steel (Group) Co., Ltd. The remaining authors declare that the research was conducted in the absence of any commercial or financial relationships that could be construed as a potential conflict of interest.

## References

1. Ukai, S.; Sakamoto, K.; Ohtsuka, S.; Yamashita, S.; Kimura, A. Alloy design and characterization of a recrystallized FeCrAl-OD cladding for accident-tolerant BWR fuels: An overview of research activity in Japan. *J. Nucl. Mater.* **2023**, *583*, 154508. [CrossRe
2. Ukai, S.; Fujiwara, M. Perspective of ODS alloys application in nuclear environments. *J. Nucl. Mater.* **2002**, *307*, 749–757. [CrossRe
3. Chung, S.; Lee, T.; Jeong, W.; Kong, B.S.; Ryu, H.J. Additive manufacturing of oxide dispersion-strengthened CoCrNi mediur entropy alloy by in situ oxide synthesis. *J. Alloys Compd.* **2023**, *965*, 171340. [CrossRef]
4. King, W.E.; Anderson, A.T.; Ferencz, R.M.; Hodge, N.E.; Kamath, C.; Khairallah, S.A.; Rubenchik, A.M. Laser powder bed fusic additive manufacturing of metals: Physics, computational, and materials challenges. *Appl. Phys. Rev.* **2015**, *2*, 041304. [CrossRe
5. Leng, B.; Ukai, S.; Sugino, Y.; Tang, Q.; Narita, T.; Hayashi, S.; Wan, F.; Ohtsuka, S.; Kaito, T. Recrystallization texture of cold-rolle oxide dispersion strengthened ferritic steel. *ISIJ Int.* **2011**, *51*, 951–957. [CrossRef]
6. Wang, J.; Liu, S.; Xu, B.; Zhang, J.; Sun, M.; Li, D. Research progress on preparation technology of oxide dispersion strengthene steel for nuclear energy. *Int. J. Extrem. Manuf.* **2021**, *3*, 032001. [CrossRef]
7. Zhong, Y.; Liu, L.; Zou, J.; Li, X.; Cui, D.; Shen, Z. Oxide dispersion strengthened stainless steel 316L with superior strength ar ductility by selective laser melting. *J. Mater. Sci. Technol.* **2020**, *42*, 97–105. [CrossRef]
8. Song, Q.S.; Zhang, Y.; Wei, Y.F.; Zhou, X.Y.; Shen, Y.F.; Zhou, Y.M.; Feng, X.M. Microstructure and mechanical performance ODS superalloys manufactured by selective laser melting. *Opt. Laser Technol.* **2021**, *144*, 107423. [CrossRef]
9. Doñate-Buendia, C.; Kürnsteiner, P.; Stern, F.; Wilms, M.B.; Streubel, R.; Kusoglu, I.M.; Tenkamp, J.; Bruder, E.; Pirch, N Barcikowski, S.; et al. Microstructure formation and mechanical properties of ODS steels built by laser additive manufacturing nanoparticle coated iron-chromium powders. *Acta Mater.* **2021**, *206*, 116566. [CrossRef]
10. Autones, L.; Aubry, P.; Ribis, J.; Leguy, H.; Legris, A.; de Carlan, Y. Assessment of ferritic ODS steels obtained by laser additiv manufacturing. *Materials* **2023**, *16*, 2397. [CrossRef] [PubMed]
11. Zhang, S.; Li, F.; Wang, Y.; Li, Z.; Zhang, S.; Kuo, Y.L.; Bai, P.; Liu, B. Study on forming process and microstructure of FeCrAl-OD alloy by selective laser melting. *J. Mater. Res. Technol.* **2025**, *35*, 1369–1376. [CrossRef]
12. Bertoli, U.S.; Wolfer, A.J.; Matthews, M.J.; Delplanque, J.P.R.; Schoenung, J.M. On the limitations of volumetric energy density as design parameter for selective laser melting. *Mater. Des.* **2017**, *113*, 331–340. [CrossRef]

Zhang, S.; Zhang, S.; Li, F.; Wang, Y.; Li, Z.; Chen, Y.; Wang, L.; Liu, B.; Bai, P. Study on defects of Zr-containing Al-Cu-Mn-Mg alloys manufactured by selective laser melting. *J. Mater. Res. Technol.* **2024**, *29*, 119–129. [CrossRef]
Yang, W.; Gan, X.; He, J. Defect Identification of 316L Stainless Steel in Selective Laser Melting Process Based on Deep Learning. *Processes* **2024**, *12*, 1054. [CrossRef]
Fardan, A.; Fazi, A.; Peng, R.L.; Mishurova, T.; Thuvander, M.; Bruno, G.; Brodin, H.; Hryha, E. Fine-Tuning Melt Pools and Microstructures: Taming Cracks in Powder Bed Fusion—Laser Beam of a non-weldable Ni-base Superalloy. *Materialia* **2024**, *34*, 102059. [CrossRef]
Liu, S.; Lee, M.; Choi, C.; Shin, K. Effect of additive manufacturing of SUS316L using selective laser melting. *J. Mater. Res. Technol.* **2023**, *24*, 9824–9833. [CrossRef]
Li, A.; Chen, Q.; Mao, J.; Wu, X.; Xin, H.; Fang, Z.; Teng, C.; Wu, L.; Tang, J. Laser powder bed fusion of oxide dispersion strengthened FeCrAl alloy: Processing and microstructural evolution. *Mater. Charact.* **2024**, *208*, 113590. [CrossRef]
Greco, S.; Gutzeit, K.; Hotz, H.; Kirsch, B.; Aurich, J.C. Selective laser melting (SLM) of AISI 316L—Impact of laser power, layer thickness, and hatch spacing on roughness, density, and microhardness at constant input energy density. *Int. J. Adv. Manuf. Technol.* **2020**, *108*, 1551–1562. [CrossRef]
Zhang, S.H.; Ukai, S.; Nishikawa, T.; Aghamiri, S.M.S.; Oono, N.; Hayashi, S. Development and characterization of g/a transformable FeCrAl-ODS alloys by cobalt addition. *J. Alloys Compd.* **2019**, *797*, 390–398. [CrossRef]
Zhang, S.; Ukai, S.; Aghamiri, S.M.S.; Oono, N.; Hayashi, S. Tensile properties of Co-added FeCrAl oxide dispersion strengthened alloy. *J. Alloys Compd.* **2021**, *852*, 156956. [CrossRef]

*materials*

MDPI

*Article*

# Laser-Induced Liquid-Phase Boron Doping of 4H-SiC

**Gunjan Kulkarni [1,2], Yahya Bougdid [2,3], Chandraika (John) Sugrim [1,4], Ranganathan Kumar [3] and Aravinda Kar [1,2,*]**

1 Electrical and Computer Engineering Department, University of Central Florida, Orlando, FL 32816, USA; gunjan.kulkarni@ucf.edu (G.K.); chandraika.sugrim.civ@us.navy.mil (C.S.)
2 Center for Research and Education in Optics and Lasers, University of Central Florida, Orlando, FL 32816, USA; yahya.bougdid@ucf.edu
3 Mechanical and Aerospace Engineering Department, University of Central Florida, Orlando, FL 32816, USA; ranganathan.kumar@ucf.edu
4 Naval Air Warfare Center—Aircraft Division (NAWCAD), Patuxent River, MD 20670, USA
* Correspondence: akar@creol.ucf.edu

**Abstract:** 4H-silicon carbide (4H-SiC) is a cornerstone for next-generation optoelectronic and power devices owing to its unparalleled thermal, electrical, and optical properties. However, its chemical inertness and low dopant diffusivity for most dopants have historically impeded effective doping. This study unveils a transformative laser-assisted boron doping technique for n-type 4H-SiC, employing a pulsed Nd:YAG laser ($\lambda$ = 1064 nm) with a liquid-phase boron precursor. By leveraging a heat-transfer model to optimize laser process parameters, we achieved dopant incorporation while preserving the crystalline integrity of the substrate. A novel optical characterization framework was developed to probe laser-induced alterations in the optical constants—refraction index (n) and attenuation index (k)—across the MIDIR spectrum ($\lambda$ = 3–5 μm). The optical properties pre- and post-laser doping were measured using Fourier-transform infrared spectrometry, and the corresponding complex refraction indices were extracted by solving a coupled system of nonlinear equations derived from single- and multi-layer absorption models. These models accounted for the angular dependence in the incident beam, enabling a more accurate determination of n and k values than conventional normal-incidence methods. Our findings indicate the formation of a boron-acceptor energy level at 0.29 eV above the 4H-SiC valence band, which corresponds to $\lambda$ = 4.3 μm. This impurity level modulated the optical response of 4H-SiC, revealing a reduction in the refraction index from 2.857 (as-received) to 2.485 (doped) at $\lambda$ = 4.3 μm. Structural characterization using Raman spectroscopy confirmed the retention of crystalline integrity post-doping, while secondary ion mass spectrometry exhibited a peak boron concentration of $1.29 \times 10^{19}$ $cm^{-3}$ and a junction depth of 450 nm. The laser-fabricated p–n junction diode demonstrated a reverse-breakdown voltage of 1668 V. These results validate the efficacy of laser doping in enabling MIDIR tunability through optical modulation and functional device fabrication in 4H-SiC. The absorption models and doping methodology together offer a comprehensive platform for paving the way for transformative advances in optoelectronics and infrared materials engineering.

**Keywords:** laser doping; silicon carbide; boron; p–n junction; refraction index

## 1. Introduction

Silicon carbide (SiC), particularly the 4H polytype, has emerged as a cornerstone material in wide-bandgap semiconductor research due to its exceptional electrical, thermal, and mechanical properties, including high breakdown electric field, wide bandgap

(3.23 eV), excellent thermal conductivity, and chemical stability [1–4]. These attributes make 4H-SiC indispensable for power electronics, high-frequency communication systems, and optoelectronic devices operating in extreme environments [5–8]. Additionally, its potential extends to advanced applications such as spin qubits in quantum computing [9], high-voltage insulation [10], and biosensing technologies [11], and its application in high-temperature and radiation-resistant environments such as aerospaces and other extreme conditions [12], reinforcing the need for precise control over its electrical and optical characteristics through advanced material processing techniques. Its exceptional combination of electrical, thermal, and mechanical properties makes SiC indispensable for high-performance applications, including gas sensors [1], power metal–oxide semiconductor field-effect transistors (MOSFETs) [3], light-emitting diodes (LEDs) [13], and p-n junction diodes [2,14].

Despite these advantages, SiC doping remains a significant challenge due to its extreme hardness and strong chemical inertness, as well as the inherently low diffusivity of most dopant species [3]. Conventional doping techniques—such as ion implantation, thermal diffusion, and epilayer doping—are often constrained by critical drawbacks, including the requirement for elevated processing temperatures, the introduction of implantation-induced lattice damage, and limitations in dopant solubility within the SiC matrix [3]. In contrast, laser doping has emerged as a compelling alternative, offering localized, non-equilibrium energy delivery that facilitates effective dopant incorporation while mitigating substrate damage. This approach supports selective-area doping, rapid thermal cycling, and the ability to surpass equilibrium solubility limits, thereby enhancing its potential for advanced semiconductor applications [15–17]. By employing high-intensity laser irradiation, dopants are introduced directly into the substrate, enabling the precise modulation of the material's electrical and optical characteristics through controlled thermal and chemical dynamics. Compared with conventional techniques, laser doping provides significant advantages, including improved dopant distribution uniformity, and fine-tuned depth control. Moreover, it enables cost-efficiencies by eliminating the necessity for prolonged high-temperature or vacuum-based treatments [13]. These attributes make laser doping particularly advantageous for the fabrication of next-generation power electronic devices, photonic components, and sensors, where precise doping control in wide-bandgap semiconductors such as SiC is critical for achieving optimal performance and reliability.

As a thermodynamically non-equilibrium process, the laser doping of SiC involves supplying high energy to a specific region of the substrate, inducing transient heating while maintaining the rest of the substrate at an ambient temperature [15]. This localized energy delivery significantly enhances dopant diffusion into the SiC lattice, positioning laser doping as an effective technique for precise, spatially controlled dopant incorporation. The technique has been successfully employed to introduce various dopant species into SiC, including gallium (Ga) [1], phosphorus (P) [2,14], nitrogen (N) [3], aluminum (Al) [3,16], chromium (Cr) [13], and boron (B) [13], enabling the development of high-performance SiC-based microelectronic devices. In contrast to conventional doping techniques that rely on solid, liquid, or gaseous precursors and often involve complex micro- or nano-fabrication steps, laser doping using solution precursors presents several distinct advantages. Solution-based precursors can provide a higher dopant concentration at the substrate surface, while offering superior safety and environmental compatibility due to their non-pyrophoric, non-toxic, and ambient-stable nature [14–16]. As a result, liquid-phase laser doping—where the SiC substrate is immersed in a dopant-containing solution and subsequently irradiated—emerges as a highly promising strategy for the selective doping of both p-type and n-type SiC substrates [14,16]. Moreover, laser doping allows for reproducible and tunable control over dopant concentration and depth profiles through adjusting laser process parameters and substrate characteristics [16]. Compared to traditional approaches, such as thermal diffusion, epilayer doping, and ion implantation, laser doping with solution precursors offers a streamlined and efficient alternative. It eliminates the need for high-temperature annealing and complex multi-step fabrication, thereby reducing the processing time, simplifying the workflow, and lowering the manufacturing costs. Collectively, this method enhances process flexibility, environmental sustainability, and precision in doping control, positioning it as a superior technique for advanced SiC device fabrication.

During the laser doping process, two key physical phenomena occur simultaneously: the laser-induced heating of the SiC substrate and the diffusion of dopant atoms into the substrate [1,2]. The laser significantly increases the temperature at the laser–substrate interaction zone, though it remains below the SiC peritectic temperature (2830 °C) [13], leading to localized melting and the subsequent re-solidification of the SiC substrate surface [1–3]. This rapid heating generates thermal gradients that promote the diffusion of dopant atoms into the substrate. As the laser continues to irradiate the surface, the dopant atoms migrate into the substrate through temperature-driven diffusion. The interaction of these processes results in the incorporation of dopant atoms into the substrate's crystal structure, which in turn alters its electrical and optical properties. Lim et al. [1] developed an uncooled MIDIR detector through utilizing the laser doping of n-type 4H-SiC with triethyl-gallium [$Ga(C_2H_5)_3$] as a Ga precursor. Their investigation unveiled the diffusion of Ga atoms into n-type 4H-SiC, creating an acceptor energy level of 0.3 eV within the SiC bandgap, corresponding to an MIDIR wavelength of $\lambda = 4.21$ μm. An analysis using FTIR spectrometry in [1] confirmed the integration of Ga atoms into the n-type 4H-SiC substrate. The exposure of Ga-doped 4H-SiC to $\lambda = 4.21$ μm caused alterations in electron density, thus affecting its transmittance, reflectance, absorptance, and refraction index. It was reported in [1] that the refraction index of SiC decreases after laser doping, reducing from 3.442 to 3.217. After doping, some dopants may occupy interstitial sites within the SiC lattice, altering the electronic structure and impacting the acceptor energy levels [1]. When dopant atoms occupy substitutional sites, they can replace silicon or carbon atoms, creating a p-type semiconductor with multiple acceptor levels. Furthermore, differences in atomic size between the dopants and the lattice atoms can introduce mechanical stress. Based on a thermal diffusion model, Bet et al. [13] calculated the diffusivity of Cr atoms at different depths within the doped SiC substrates as a function of temperature induced by laser. They reported a maximum diffusivity of Cr atoms in SiC of $6.9 \times 10^{-12}$ $cm^2/s$ at 3046 K for 6H-SiC and $4.61 \times 10^{-10}$ $cm^2/s$ at 2898 K for 4H-SiC substrates. Thermal doping techniques are inherently limited by the solid solubility limits of the material, while ion implantation can surpass these limits but often causes significant damage to the implanted wafer [13]. In contrast, laser doping, a non-equilibrium process, enables the solubility limits to be exceeded with minimal damage. This extended solubility has been observed for dopants such as Al, N and B, as addressed in [1,3,13]. Several factors contribute to this increased solubility: (i) surface interactions reduce the stress caused by atomic size mismatches between C, Si, and dopant atoms; (ii) atomic reconstruction near the surface generates stress fields; and (iii) the pulsed laser's rapid heating and cooling cycles trap dopants by causing the lattice to expand and contract quickly. The combined effects of impurity-induced stress and spatial variations in surface and thermal stress significantly lower the energy required for dopants to occupy specific lattice sites in the SiC substrate. Barkby et al. [17] introduced a single-step method called high-pressure gas-immersion excimer laser doping (HP-GIELD) for creating nitrogen-hyperdoped silicon. This technique offers ultrafast processing, scalability, precise control, and excellent reproducibility. Using HP-GIELD, they successfully produced nitrogen-hyperdoped silicon in nitrogen-rich environments. By optimizing the process parameters, they achieved nitrogen concentrations exceeding $3.01 \times 10^{21}$ $cm^{-3}$, which is 12 times higher than the previously reported maximum dopant concentration and six orders of magnitude higher than the solid solubility limit [17]. Laser doping enabled the formation of interconnects on insulating SiC surfaces without requiring a metallization process [18,19]. Consequently, it presents a potential alternative to conventional doping methods [3], especially for wide-bandgap semiconductor materials.

Excimer lasers, such as ArF (λ = 193 nm), KrF (λ = 248 nm), and XeCl (λ = 308 nm), have been widely explored for doping SiC substrates. However, these lasers induce strong photon–material interactions, leading to excessive localized heating and surface ablation. This can compromise the structural integrity of SiC and degrade its electrical properties, making excimer lasers less applicable for doping [2,14,16]. In contrast, the Nd:YAG laser (λ = 1064 nm) offers several advantages over excimer lasers for doping. Its lower photon energy and deeper dopant penetration enables higher surface temperatures and deeper heat conduction into the substrate, which is beneficial for efficient dopant incorporation [3]. Additionally, the longer irradiation time associated with the Nd:YAG laser helps maintain an elevated substrate temperature, thereby enhancing the diffusion of boron atoms into SiC. Previous studies have employed the Nd:YAG laser for doping with conventional precursors [1,13], but these approaches have struggled to achieve high and deep dopant incorporation. To address these limitations, this work investigates the use of an NIR pulsed Nd:YAG laser (λ = 1064 nm) in combination with a solution precursor. This approach has the potential to significantly enhance dopant penetration and improve process efficiency, offering a more effective method for the laser doping of SiC. Aqueous boric acid ($H_3BO_3$) solution was chosen as the boron precursor in the current study. A heat-transfer model was used to guide the selection of appropriate laser process parameters for doping crystalline 4H-SiC substrates, ensuring that phase transformations are avoided. FTIR spectrometry was conducted within the MIDIR range to analyze the changes in the optical properties, specifically the reflectance and absorptance of the as-received and B-doped 4H-SiC substrates. The absorption peak at λ = 4.3 μm, corresponding to the acceptor level at 0.29 eV created by boron atoms within the 4H-SiC bandgap, was identified. The absorption models developed in the current study were presented to determine the refraction and attenuation indices before and after doping. The results obtained from the absorption models were compared with those obtained using Fresnel's method [15]. Additionally, the impact of variations in the incident angle on the measured reflectance and their effects on the optical properties are reported. The absorption models outlined in this paper are essential for precisely assessing the optical properties of SiC in laser doping technologies.

In prior research [15], we demonstrated the feasibility of laser doping 4H-SiC using a solution-based precursor and analyzed its influence on the SiC's optical properties using Fourier-transform infrared (FTIR) spectrometry at normal incidence. However, this earlier study was limited in scope, focusing solely on optical responses and employing Fresnel's equation without independent validation for determination of the refraction index. Furthermore, no electrical characterization or comparative assessment against conventional doping techniques was undertaken. The present study significantly advances beyond our previous work by integrating both optical and electrical characterizations of laser-doped 4H-SiC, thereby offering a more holistic evaluation of the doping process. Specifically, we introduced robust, angle-resolved absorption models to accurately extract the complex refraction index (n + ik) of as-received and boron-doped SiC, validated against Fresnel's method. These models allow for the precise determination of optical constants under varying incidence angles, providing a more comprehensive understanding of doping-induced modifications. Crucially, this work includes a detailed electrical characterization of a laser-fabricated p–n junction diode, demonstrating the functional viability of the laser doping approach. The diode exhibited a reverse-breakdown voltage of 1668 V, and temperature-dependent resistivity measurements confirmed the electrical activation of the boron dopant. Secondary ion mass spectrometry (SIMS) analysis revealed a peak boron concentration of $1.29 \times 10^{19}$ cm$^{-3}$ and a junction depth of 450 nm, indicating effective dopant diffusion. This study not only refines the methodology for optical property extraction in laser-doped SiC but also validates the efficacy of the doping process through the electrical performance metrics of fabricated devices. By addressing both optical and electrical domains, and comparing the developed technique with established doping methods, this work provides a comprehensive platform for advancing SiC-based optoelectronic and power device technologies.

## 2. Process Parameters Selection

To enable the effective laser doping of 4H-SiC, it is critical to optimize laser process parameters to enhance dopant diffusion while preventing thermal damage or phase transitions. This requires precise control of the laser power, irradiation duration, and optical absorption to ensure that the surface temperature stays below the peritectic point of SiC (2830 °C) [13]. To support this analysis, the optical properties of as-received n-type 4H-SiC substrate at λ = 1064 nm were characterized using a Lambda 365+ UV–Vis spectrophotometer (Perkin Elmer, Inc., Waltham, MA, USA). The measured transmittance (T = 18.05%), reflectance (R = 35.13%), and absorptance (A = 46.79%) at λ = 1064 nm provide critical input for estimating the laser process parameters corresponding to the optimal theoretical doping temperature. In particular, the absorptance of the SiC substrate directly governs the energy coupling at the laser–substrate interaction zone, thereby strongly influencing the peak temperature reached during irradiation [15].

The temperature at the laser–substrate interaction zone can be estimated by incorporating parameters such as laser power, spot size, the absorptance of 4H-SiC at λ = 1064 nm, the thermal conductivity, the thermal diffusivity, and the laser–substrate interaction time [15]. A comprehensive heat transfer model was employed to compute the doping temperature as a function of laser power and substrate scanning velocity, using the thermophysical and optical properties of the SiC substrate [20]. The calculated surface temperature, shown in Figure 1a, varies with both laser power and scanning speed. The results demonstrate that lower scanning velocities facilitate increased thermal exposure, thereby reducing thermal stress while enabling efficient dopant diffusion without inducing structural damage or phase transitions. To ensure substrate integrity during laser doping, it is also critical to evaluate the heating rate, defined as the temporal rate of temperature rise at the laser–substrate interface, $\left(\frac{dT(0,t)}{dt}\right)$. The heating rate of the substrate at any time $t = t_{int}$, where $t_{int}$ is the interaction time, is determined by Equation (1). In Equation (1), T represents the temperature, A is the absorptance of SiC at λ = 1064 nm, P is the irradiation laser power, κ is the thermal conductivity, a is the area of laser–SiC substrate, κ is the thermal diffusivity, $v$ is the scanning velocity, and d is the diameter of the laser spot [20,21]:

$$\frac{dT}{dt} = \left(\frac{AP}{\kappa a}\right)\sqrt{\frac{\kappa v}{\pi d}} \quad (1)$$

The overall scenario, which encompasses both spatial and temporal changes in the substrate temperature during the doping process, can be described using Equation (2), obtained from [20,21]:

$$T(z,t) = \left\{\frac{2Pe^{-\left[\frac{z^2}{4\kappa t}\right]}}{\kappa a}\sqrt{\frac{\kappa t}{\pi}}\right\} - \left\{\frac{Pz}{a}\operatorname{erfc}\left[\frac{z}{2\sqrt{\kappa t}}\right]\right\} \quad (2)$$

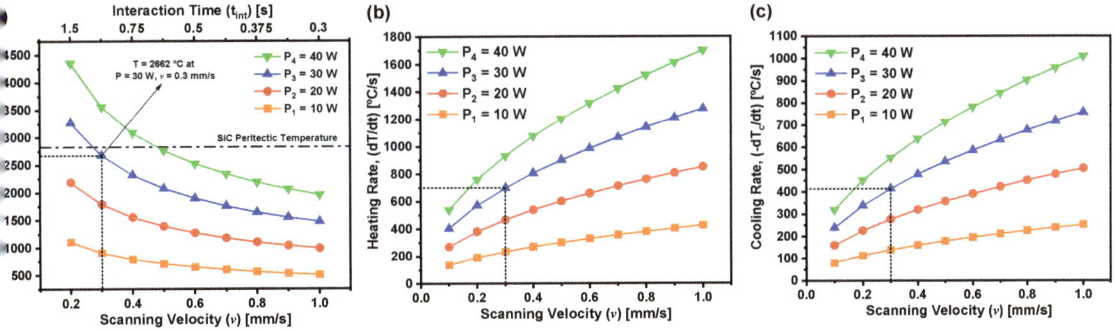


**Figure 1.** (**a**) Temperature at the laser–SiC substrate interaction zone, (**b**) heating rate, and (**c**) cooling rate as a function of scanning velocity for various laser powers.

Here, the vertical axis, denoted as z, is oriented downwards and originates from the center of the laser beam. By considering $T(z, t_{int})$ in Equation (2) as the initial condition, the solution to the one-dimensional transient heat conduction equation (Equation (1)) depicts the temperature distribution, $T_c(z, t)$, within the material during the cooling phase after the laser beam has heated the interaction zone, at which point doping can be induced. To obtain Equation (3), the derivative of $T_c(z, t)$ with respect to time is evaluated at z = 0 and $t = t_{int}$; this signifies the cooling rate at the specific $t_{int}$ [20,21].

$$-\frac{dT_c}{dt} = \left(\frac{0.293P}{\kappa a}\right)\sqrt{\frac{\kappa \upsilon}{\pi d}} \tag{3}$$

The heat-transfer model was previously validated using a simplified approach [21] to predict laser-induced temperature profiles during the sintering of $TiO_2$ nanoparticles (NPs) on quartz substrates. The model accurately forecasted the anatase-to-rutile phase transformation, predicting a temperature range of 550–600 °C under laser irradiation at 3.55 W and a scanning speed of 2 mm/s. These conditions corresponded to the experimentally observed phase change, as evidenced by the formation of rutile-phase $TiO_2$ and the characteristic brown coloration shown in Figure 8g of [21]. This validation established the model's effectiveness in reliably correlating laser process parameters with phase transitions.

Building on the validated framework, this study employs the heat-transfer model to optimize boron doping in 4H-SiC by predicting the substrate's thermal response under various laser powers and scanning velocities, minimizing experimental iterations and preventing thermal degradation. Laser-induced heating facilitates boron incorporation through localized thermal activation, focused around a 300 µm laser spot. Crucially, the model ensures that the doping temperature remains below the SiC peritectic temperature (2830 °C), thereby preventing crystalline phase transformations. Among the tested conditions, laser powers of 10 W and 20 W yielded estimated surface temperatures of 910 °C and 1794 °C, respectively, which were insufficient for effective boron diffusion. A laser power of 30 W was found to be optimal, producing a localized temperature of ~2662 °C—this is adequate to allow for the substitutional diffusion of boron atoms into the SiC lattice (Figure 1b). Additionally, the model predicted that lower scanning velocities reduce cooling rates, minimizing thermal gradients that could induce mechanical stress. For instance, at 40 W and a scanning speed of 1 mm/s, the model indicated a rapid thermal rise followed by a steep cooling, increasing the risk of substrate cracking (Figure 1c). As a result, a power–speed combination of 30 W and 0.3 mm/s was selected to ensure sufficient thermal energy for doping, while enabling controlled cooling to prevent substrate breakage. By establishing a quantitative link between laser parameters and thermal behavior, the model facilitates precision control over dopant incorporation and structural integrity. This predictive capability enhances process reliability and reduces the need for extensive empirical tuning, underscoring the model's value in advanced material processing.

## 3. Experimental Procedure

### 3.1. SiC Substrate Cleaning Procedure and Preparation of Dopant Solution

An n-type 4H-SiC wafer with a thickness of D = 350 μm and nitrogen dopant concentration ($N_D$) of ~$10^{16}$ $cm^{-3}$ was procured from II-VI Inc. (Pine Brook, NJ, USA) and diced to produce 10 × 10 $mm^2$ substrates. Initially, the SiC substrates were scrubbed with soap and rinsed thoroughly with deionized (DI) water. They were then immersed in a 1:1 mixture of 98% sulfuric acid ($H_2SO_4$) and 30 wt.% hydrogen peroxide ($H_2O_2$) for 15 min, followed by another rinse with DI water. Next, the substrates were treated with a 6:1 buffered oxide etchant (BOE) for 10 min to eliminate any native silicon dioxide ($SiO_2$) on their surfaces. After rinsing with DI water, they were cleaned with acetone and methanol, rinsed again with DI water, and finally dried with nitrogen ($N_2$) gas.

The dopant solution was formulated using boric acid ($H_3BO_3$) powder as the precursor. All chemicals used in this study were purchased from Sigma-Aldrich, Inc. (Burlington, MA, USA). An aqueous boric acid solution with a concentration of 4.63 wt.% (equivalent to 0.75 M) was prepared, yielding a boron concentration of 4.52 × $10^{20}$ $cm^{-3}$. This formulation deliberately surpasses the established solid solubility limit of boron in SiC (2.5 × $10^{20}$ $cm^{-3}$) [13], ensuring a high density of boron dopants is supplied to the substrate surface before laser-assisted doping.

### 3.2. Optical Setup for Laser Doping

The schematic of the optical setup for laser doping is shown in Figure 2. The Nd:YAG laser (Lee Laser Inc., Model: LEE-8150MQ, Orlando, FL, USA) employed, with a 170 ns pulse width and a 30 kHz repetition rate, enabled prolonged irradiation and maintained an elevated substrate temperature, thereby facilitating the diffusion of boron atoms into SiC. The Gaussian beam from the laser system was transformed into a cylindrical annular beam using an arrangement of axicon and biconvex lenses. It was then focused into a hollow laser cone using a gold-coated parabolic mirror, resulting in a Gaussian–Bessel beam of ~300 μm in diameter. The interaction between the laser beam, dopant solution, and substrate occurred at the apex of the cone. The laser beam was scanned across a 6 × 6 $mm^2$ area of the SiC substrate using a motorized 3D stage. Laser irradiation can induce the formation of additional vacancies in SiC as a result of stress mechanisms such as thermal expansion, shock wave generation, and vibrational excitation. Thus, the 4H-SiC substrate was irradiated at ambient temperature with 30 W power and 0.3 mm/s scanning velocity to create vacancies by out-diffusing nitrogen atoms from the SiC surface and to promote boron diffusion. Doping was carried out on the Si-face of the substrate with a 1 sec interaction time per scan. Subsequently, the substrate was immersed in the dopant solution, where the thickness of the solution layer above the SiC surface was carefully adjusted to ~0.1 mm to avoid interference with the laser doping process due to absorption [22].

During the laser doping process, the 4H-SiC substrate was submerged in the dopant solution and subjected to laser irradiation. The localized heating induced by the laser resulted in the thermal decomposition of the dopant solution at the laser-heated spot, generating boron atoms that subsequently diffused into SiC. According to the heat-transfer model described in Section 2, the doping temperature briefly reached 2662 °C at the applied power of 30 W and 0.3 mm/s speed, remaining below the SiC peritectic threshold (2830 °C). The rapid heating facilitated boron diffusion while the laser pulse was ON, and when the laser pulse was OFF, the substrate underwent rapid cooling, effectively immobilizing the boron dopant atoms within the SiC lattice. Post-doping, the substrate was cleaned with an aqueous 45 wt.% potassium hydroxide (KOH) solution, followed by rinsing with acetone, methanol, and water, and was then dried with $N_2$ gas, respectively.

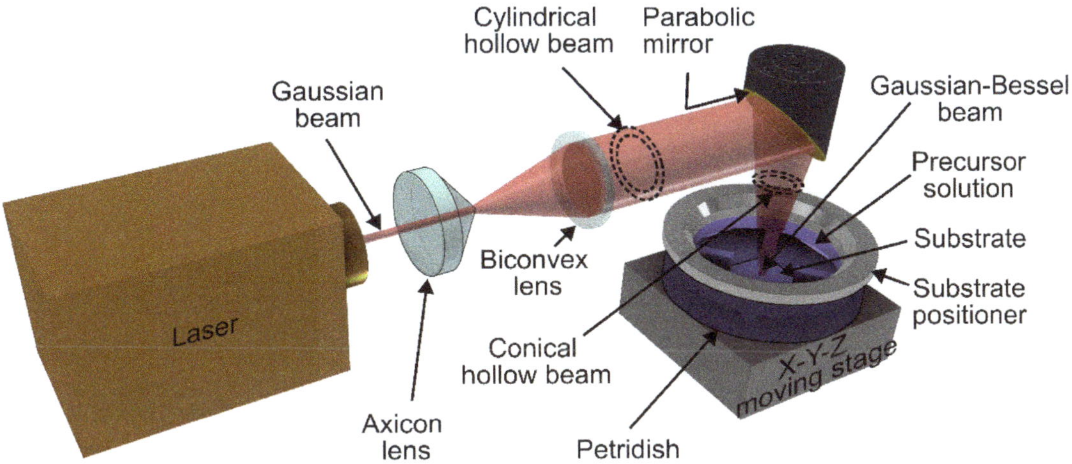


**Figure 2.** Diagrammatic representation of the optical configuration employed for laser doping of the 4H-SiC substrate.

## 4. Absorption Models for the Determination of Refraction and Attenuation Indices Pre- and Post-Laser Doping

Hecht [23] presented the electromagnetic theory of light reflection and transmission through both single and multilayer films composed of homogeneous, non-magnetic, and linear isotropic media. The transfer matrix method in [23] for calculating the reflection and transmission coefficients incorporates both the real and imaginary components of the refraction index. This method proves particularly effective for analyzing variations in reflection coefficients due to differing refraction indices across multiple layers. Building on Hecht's foundational work, we developed absorption models to calculate the real (refraction index, n) and imaginary (attenuation index, k) components of the complex refraction index of SiC substrates, both before and after laser doping. In this section, we present the single- and multi-layer absorption models used to determine n and k for the as-received (undoped) and doped regions of the SiC substrates. These models utilize FTIR spectrometry measurements at various incident angles to evaluate the optical properties of both the undoped and doped regions of the SiC substrate.

In Figure 3, $n_0$ denotes the refraction index of air (media, m = 0 and m = 3), $n_1$ and $k_1$ denote the refraction and attenuation indices of the doped region (medium, m = 1) of the substrate, and $n_2$ and $k_2$ represent the refraction and attenuation indices for the undoped region (medium, m = 2). The angles $\theta_0$ and $\theta_3$ are the known real angles from experimental conditions, while $\tilde{\theta}_1$ and $\tilde{\theta}_2$ are complex angles for oblique incidence. Complex angles represent the effective angles of refraction or transmission in lossy materials, combining phase rotation (real part) and attenuation due to absorption (imaginary part) [24]. The real part indicates how the wave's direction changes, while the imaginary part describes the reduction in amplitude from absorption. This concept is crucial for understanding wave behavior in materials such as doped 4H-SiC substrates, where both refracted angle and attenuation must be considered.

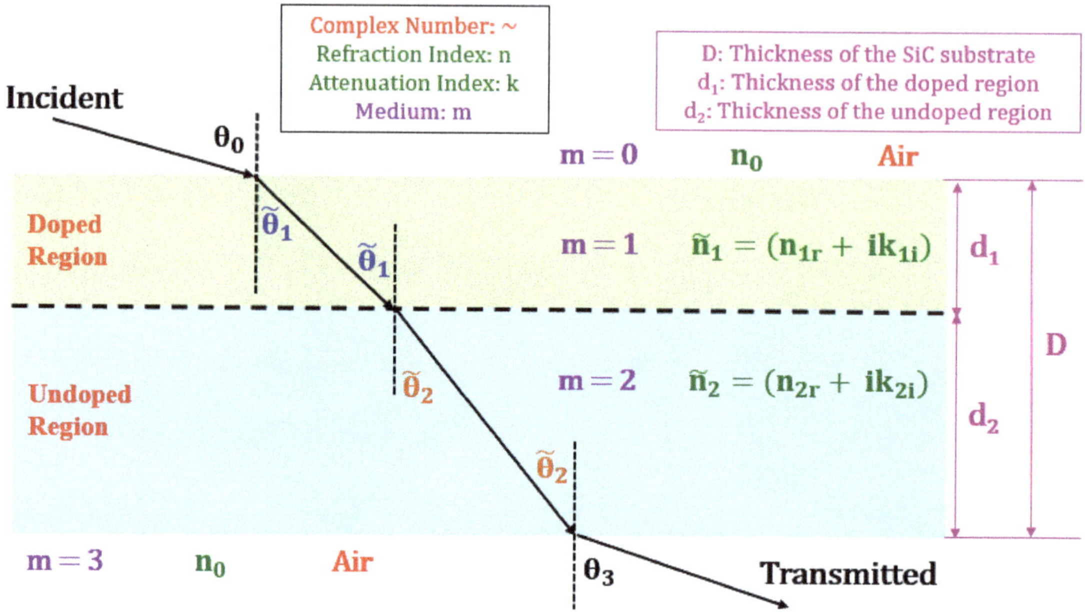


**Figure 3.** Schematic of light propagation through doped and undoped regions of the SiC substrate.

*4.1. Single-Layer Absorption Model for Determining $n_2$ and $k_2$ for the Undoped SiC Region*

The optical properties of the undoped SiC region were determined using the absorption model outlined below. The relationship between the reflection coefficient (r), wavelength (λ) and angle of incidence (θ) for both perpendicular (Equation (4)) and parallel polarizations (Equation (5)) can be expressed as follows [25]:

$$r_s = \frac{a^2 + b^2 - 2a\cos\theta + \cos^2\theta}{a^2 + b^2 + 2a\cos\theta + \cos^2\theta} \tag{4}$$

$$r_p = \frac{a^2 + b^2 - 2a\sin\theta\tan\theta + \sin^2\theta\tan^2\theta}{a^2 + b^2 + 2a\sin\theta\tan\theta + \sin^2\theta\tan^2\theta}\,[r_s] \tag{5}$$

where

$$a = \frac{1}{\sqrt{2}}\left\{\left[\left(n_2^2 - k_2^2 - \sin^2\theta\right)^2 + 4\,n_2^2k_2^2\right]^{\frac{1}{2}} + n_2^2 - k_2^2 - \sin^2\theta\right\}^{\frac{1}{2}} \tag{6}$$

$$b = \frac{1}{\sqrt{2}}\left\{\left[\left(n_2^2 - k_2^2 - \sin^2\theta\right)^2 + 4\,n_2^2k_2^2\right]^{\frac{1}{2}} - n_2^2 + k_2^2 + \sin^2\theta\right\}^{\frac{1}{2}} \tag{7}$$

The reflectance of the undoped SiC region ($R_U$) is determined by averaging the perpendicular and parallel polarization components of r, and is expressed as follows:

$$R_U = \frac{r_s + r_p}{2} \tag{8}$$

The simplified expression for $R_U$ is obtained by expressing it in terms of $n_2$ and $k_2$, as follows:

$$2R_U = \left[\frac{\left[\left(n_2^2-k_2^2-\sin^2\theta\right)^2+4n_2^2\,k_2^2\right]^{½}-\left(\sqrt{2}\left[\left[\left(n_2^2-k_2^2-\sin^2\theta\right)^2+4n_2^2\,k_2^2\right]^{½}+n_2^2-k_2^2-\sin^2\theta\right]^{\frac{1}{2}}\times\cos\theta\right)+\cos^2\theta}{\left[\left(n_2^2-k_2^2-\sin^2\theta\right)^2+4n_2^2\,k_2^2\right]^{½}+\left(\sqrt{2}\left[\left[\left(n_2^2-k_2^2-\sin^2\theta\right)^2+4n_2^2\,k_2^2\right]^{½}+n_2^2-k_2^2-\sin^2\theta\right]^{\frac{1}{2}}\times\cos\theta\right)+\cos^2\theta}\right]\times$$
$$\left\{\frac{\left[\left(n_2^2-k_2^2-\sin^2\theta\right)^2+4n_2^2\,k_2^2\right]^{½}-\left(\sqrt{2}\left[\left[\left(n_2^2-k_2^2-\sin^2\theta\right)^2+4n_2^2\,k_2^2\right]^{½}+n_2^2-k_2^2-\sin^2\theta\right]^{\frac{1}{2}}\times\sin\theta\,\tan\theta\right)+\sin^2\theta\,\tan^2\theta}{\left[\left(n_2^2-k_2^2-\sin^2\theta\right)^2+4n_2^2\,k_2^2\right]^{½}+\left(\sqrt{2}\left[\left[\left(n_2^2-k_2^2-\sin^2\theta\right)^2+4n_2^2\,k_2^2\right]^{½}+n_2^2-k_2^2-\sin^2\theta\right]^{\frac{1}{2}}\times\sin\theta\,\tan\theta\right)+\sin^2\theta\,\tan^2\theta}+1\right\} \quad (9)$$

Since the parameters $R_U$, $\theta$, are either known or measured for the respective wavelength, they can be substituted into Equation (9) to yield equations involving $n_2$ and $k_2$. By solving these equations simultaneously, the values of $n_2$ and $k_2$ were determined.

### *4.2. Multi-Layer Absorption Model for Determining $n_1$ and $k_1$ for the Doped SiC Region*

To determine $n_1$ and $k_1$ for the doped SiC region, where $\tilde{n}_1 = n_{1r} + ik_{1i}$, we apply Snell's law to the schematic in Figure 3, yielding the following:

$$n_0 \sin\theta_0 = \tilde{n}_1 \sin\tilde{\theta}_1 = \tilde{n}_2 \sin\tilde{\theta}_2 = n_0 \sin\theta_3 \quad (10)$$

As per Equation (10), $\theta_0 = \theta_3$.

The perpendicular and parallel components of r for each interface shown in Figure 3 can be expressed as follows [23]:

$$\tilde{r}_s^{(ab)} = \frac{\tilde{n}_a \cos\tilde{\theta}_a - \tilde{n}_b \cos\tilde{\theta}_b}{\tilde{n}_a \cos\tilde{\theta}_a + \tilde{n}_b \cos\tilde{\theta}_b} \quad (11)$$

$$\tilde{r}_p^{(ab)} = \frac{\tilde{n}_b \cos\tilde{\theta}_a - \tilde{n}_a \cos\tilde{\theta}_b}{\tilde{n}_b \cos\tilde{\theta}_a + \tilde{n}_a \cos\tilde{\theta}_b} \quad (12)$$

In Equations (11) and (12), the subscript 'a' refers to the incident medium, while 'b' denotes the transmitted medium. The superscript 'ab' represents the interface between two media, such as air–doped region (interface: 01), doped–undoped region (interface: 12), and undoped region–air (interface: 23), as illustrated in Figure 3.

The propagation factor for plane waves is provided by Equation (13) [23]:

$$\tilde{\delta}_m = \frac{2\pi\,\tilde{n}_m\,d_m \cos\tilde{\theta}_m}{\lambda_0} \quad (13)$$

Here, $\lambda_0 = 4.3$ μm, with 'm' representing either 1 or 2 to indicate media 1 or 2, respectively. 'd' denotes the thickness, and 'θ' is the incident angle for the corresponding medium. The thickness of medium–1 (doped SiC region), denoted as $d_1$ is shown in Figure 3. The value of $d_1$ was measured to be 450 nm using Secondary Ion Mass Spectrometry (SIMS) (see Section 5.3).

The perpendicular and parallel components of r across the entire SiC substrate can be expressed as follows [23]:

$$\tilde{r}_s = \frac{\left(\tilde{r}_s^{(01)} + \tilde{r}_s^{(12)} e^{i2\tilde{\delta}_1}\right) + \left(\tilde{r}_s^{(01)}\tilde{r}_s^{(12)} + e^{i2\tilde{\delta}_1}\right)\tilde{r}_s^{(23)} e^{i2\tilde{\delta}_2}}{\left(1 + \tilde{r}_s^{(01)}\tilde{r}_s^{(12)} e^{i2\tilde{\delta}_1}\right) + \left(\tilde{r}_s^{(12)} + \tilde{r}_s^{(01)} e^{i2\tilde{\delta}_1}\right)\tilde{r}_s^{(23)} e^{i2\tilde{\delta}_2}} \quad (14)$$

$$\tilde{r}_p = \frac{\left(\tilde{r}_p^{(01)} + \tilde{r}_p^{(12)} e^{i2\tilde{\delta}_1}\right) + \left(\tilde{r}_p^{(01)}\tilde{r}_p^{(12)} + e^{i2\tilde{\delta}_1}\right)\tilde{r}_p^{(23)} e^{i2\tilde{\delta}_2}}{\left(1 + \tilde{r}_p^{(01)}\tilde{r}_p^{(12)} e^{i2\tilde{\delta}_1}\right) + \left(\tilde{r}_p^{(12)} + \tilde{r}_p^{(01)} e^{i2\tilde{\delta}_1}\right)\tilde{r}_p^{(23)} e^{i2\tilde{\delta}_2}} \tag{15}$$

Starting from Equations (11) and (12), the perpendicular and parallel components of reflectance, denoted as $R_s$ and $R_p$, respectively, can be determined as follows [23]:

$$R_s = \tilde{r}_s \tilde{r}_s^* \tag{16}$$

$$R_p = \tilde{r}_p \tilde{r}_p^* \tag{17}$$

where $\tilde{r}_s^*$ and $\tilde{r}_p^*$ are complex conjugates of $\tilde{r}_s$ and $\tilde{r}_p$, respectively.

The reflectance of the doped SiC region ($R_D$) is the average of both $R_s$ and $R_p$, and can be expressed as follows:

$$R_D = \frac{R_s + R_p}{2} \tag{18}$$

Given that the parameters $R_D$, $\theta$, $d_1$, $d_2$ and $\lambda_0$ are either known or measured, they can be substituted into Equation (18) to yield equations involving $n_1$ and $k_1$. By solving these equations simultaneously, the values of $n_1$ and $k_1$ were determined. Detailed derivations for $n_1$ and $k_1$ are provided in the Supplementary Materials.

## 5. Results and Discussion

### *5.1. Raman Spectra Analysis*

Figure 4 presents the Raman spectra of as-received and boron-doped 4H-SiC substrates obtained using LabRAM HR Evolution Nano (Horiba Instruments Inc., Irvine, CA, USA), revealing distinct modifications in vibrational characteristics resulting from boron incorporation. In the as-received 4H-SiC, the primary Raman-active modes appear at ~772 $cm^{-1}$ ($E_{2g}$) and ~960 $cm^{-1}$ ($A_{1g}$), corresponding to in-plane and out-of-plane vibrations of Si–C bonds, respectively, consistent with the 4H polytype's crystalline symmetry [26]. Following boron doping, these modes exhibited notable blue shifts (shift to higher wavenumbers) to 781.355 $cm^{-1}$ and 968.645 $cm^{-1}$, respectively. These spectral shifts are indicative of the compressive lattice strain introduced by boron atoms being substituted into the SiC crystal lattice, which alters local bonding configurations by reducing interatomic spacing and increasing bond stiffness, thereby enhancing phonon frequencies [27]. In addition to peak shifts, a reduction in Raman intensity is observed in the doped sample. This attenuation is attributed to the enhanced phonon scattering caused by defect generation and plasmon–phonon coupling, where interactions between free carriers and phonons suppress the Raman scattering efficiency [27]. Moreover, the appearance of broad, additional features in the 1250–1700 $cm^{-1}$ range suggests the presence of disorder-induced vibrational modes, potentially linked to carbon-related phases. Specifically, the observed bands near ~1350 $cm^{-1}$ and ~1580 $cm^{-1}$ correspond to the D-band and G-band of $sp^2$-bonded carbon, respectively [28]. These features may arise from local structural disorder, non-stoichiometric carbon clustering, or the formation of graphitic-like domains during the high-temperature laser doping process. These spectral variations highlight the substantial impact of boron doping on the lattice dynamics and structural integrity of 4H-SiC. The observed strain effects, defect signatures, and potential carbon phase formation offer valuable insight into dopant incorporation mechanisms and their influence on the material's electronic and optoelectronic properties.

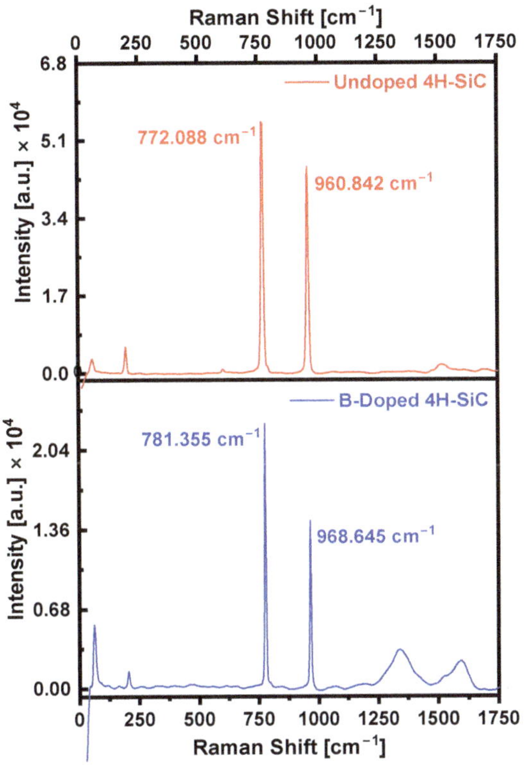


**Figure 4.** Raman spectra of as-received (red) and boron-doped (blue) 4H-SiC substrates, highlighting the effects of boron incorporation on vibrational modes.

*5.2. Temperature-Dependent Resistivity and Ionization Energy of Boron in 4H-SiC*

The resistivity (ρ) of boron-doped 4H-SiC can be expressed as $\rho = \rho_0 \bullet e^{\left(\frac{E_i}{KT}\right)}$ [29], where $\rho_0$ is the pre-exponential factor, T is the absolute temperature, k is the Boltzmann constant, and $E_i$ is the ionization energy of the boron dopant. Figure 5 illustrates the temperature-dependent resistivity of boron-doped 4H-SiC obtained using P2010A2 Four Point Probe (Ossila Ltd., Sheffield, UK). The slope (m) of the linear fit was determined to be ~3453, which corresponds to $E_i$ ~298 meV, as per the relation $E_i = m \bullet k$. The ionization energy in the current study is in good agreement with the previously reported ionization energies for boron acceptors in equilibrium-doped 4H-SiC, which typically range from ~280 to 390 meV [30,31]. However, the extracted value here is particularly significant because doping was achieved via pulsed laser, a non-equilibrium technique. Such laser-induced processes can locally modify the lattice structure through rapid thermal gradients, generate residual point defects, and possibly promote impurity band formation at high dopant concentrations. These effects are known to alter the local electronic structure and reduce the effective ionization barrier for hole-generation from acceptor states. The extracted ionization energy, compared to undisturbed bulk values, suggests the enhanced electrical activation of boron. This implies a higher ionization fraction (f) at room temperature, thereby increasing the availability of free holes for conduction. Such behavior is especially advantageous for p-type conductivity in SiC and can significantly improve the performance of boron-doped layers in p–n junctions, photodetectors, and other SiC-based optoelectronic or power devices.

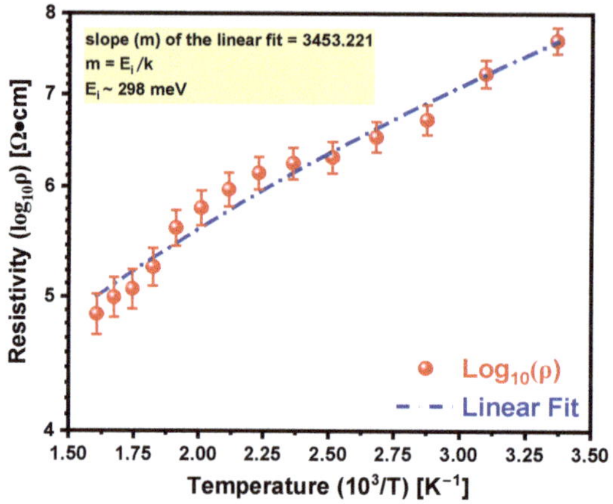


**Figure 5.** Arrhenius plot of resistivity for boron-doped 4H-SiC. Red circles with standard deviations represent experimentally measured resistivity values across the 300–600 K temperature range, while the dashed blue line shows the linear fit used to determine the slope. The linear behavior of $\log_{10}(\rho)$ vs. $10^3/T$ confirms the thermally activated conduction and enables the extraction of the ionization energy of the boron dopant.

### *5.3. Secondary Ion Mass Spectrometry (SIMS) Analysis*

Figure 6a presents the SIMS depth profile of boron in laser-doped 4H-SiC obtained using PHI Adept 1010 Dynamic SIMS System, showing both the total acceptor concentration ($N_A$) and the ionized acceptor concentration ($N_A^-$) as functions of depth. The original SIMS data, recorded in arbitrary units, were quantitatively calibrated using a boron-implanted reference sample fabricated through implanting boron ions into n-type 4H-SiC at a dose of $5 \times 10^{15}$ ions/cm$^2$ at 100 keV, enabling the conversion of raw SIMS counts to absolute concentrations in units of cm$^{-3}$. A peak boron concentration of $1.29 \times 10^{19}$ cm$^{-3}$ was observed at the surface, followed by a sharp decline to $4.72 \times 10^{17}$ cm$^{-3}$ within the first ~200 nm. This rapid drop was attributed to the substitutional diffusion driven by a steep laser-induced thermal gradient. From 200 to 450 nm, the concentration stabilizes, forming a quasi-plateau, before gradually decreasing, consistent with the knock-on diffusion effects influenced by depth-dependent cooling. $N_A^-$ as a function of depth was determined using the relation, $N_A^- = \left(N_A \bullet e^{\frac{-E_i}{kT}}\right)$ [29,32]. Figure 6a shows that $N_A^-$ is slightly lower than $N_A$ across the depth profile. This disparity highlights the limited electrical activation of boron in 4H-SiC, consistent with prior findings [1,15,18].

Figure 6b presents the measured total acceptor concentration ($N_A$) profile in laser-doped 4H-SiC (red curve), obtained via SIMS, alongside a theoretical fit (purple curve) using the complementary error function (erfc) solution to the one-dimensional diffusion equation. The fitting curve employs the analytical solution, $C(x) = N_A \bullet \text{erfc}\left(\frac{x}{2\sqrt{D_T \bullet t_{int}}}\right)$ [33], where $D_T$ is the diffusion coefficient and $t_{int} = 1$ s. The fit analysis yielded a diffusion coefficient of $D_T = 9.83 \times 10^{-8}$ cm$^2$/s, which is several orders of magnitude higher than the values previously reported for boron diffusion in SiC under equilibrium conditions [30,34]. Roccaforte et al. [30] reported boron diffusivities in the range of $5.5 \times 10^{-15}$ cm$^2$/s at 1800 °C to $10^{-11}$ cm$^2$/s at 2500 °C, while Kubiak et al. [34] documented a lower diffusivity of ~$5 \times 10^{-9}$ cm$^2$/s for boron in 4H-SiC due to its strong covalent bonding and limited diffusion pathways. The markedly enhanced diffusivity observed in this work was attributed to the non-equilibrium nature of pulsed laser doping, which likely generated transient high-temperature conditions, induced point defects, and possibly led to localized melting or amorphization. These effects facilitated rapid boron incorporation and transport, enabling the efficient formation of shallow, highly doped p-type regions, thereby demonstrating the potential of laser doping for advanced SiC device applications.

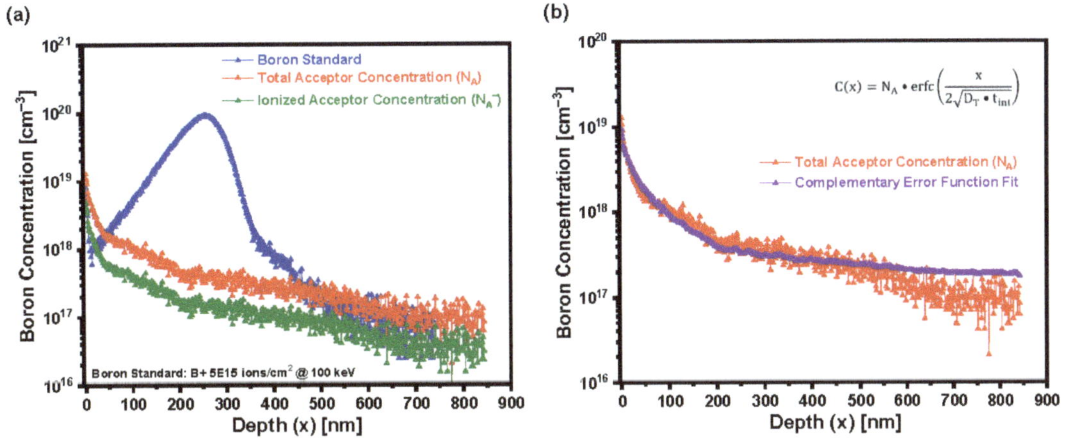


**Figure 6.** (**a**) SIMS profiles of the laser-doped 4H-SiC substrate, displaying the total and effective acceptor concentrations alongside the boron standard as a function of depth (x). (**b**) The complementary error function (erfc) fit for the total acceptor concentration, yielding a diffusion coefficient of $D_T = 9.83 \times 10^{-8}$ cm$^2$/s.

Figure 7 shows the temperature-dependent ionization fraction (f) of boron acceptors in 4H-SiC as a function of temperature (300–3000 K), calculated using $E_i$ ~298 meV and acceptor degeneracy factor, $g_p = 4$ [31]. The relationship is given by $f = \left\{1 + \left(g_p \bullet e^{\frac{E_i}{kT}}\right)\right\}^{-1}$ [29]. The ionization fraction increases steadily with temperature, rising from nearly 0 at 300 K to ~0.164 at 3000 K. This trend reflects the deep acceptor nature of boron in 4H-SiC, where a significant amount of thermal energy is required to activate dopants. Even under elevated thermal conditions (e.g., 2000 K), a comparatively small fraction of boron atoms are ionized. These results highlight the importance of accounting for incomplete ionization when estimating the electrically active dopant concentrations $\left(N_A^- = f \bullet N_A\right)$ and when modeling temperature-dependent electrical or optical behavior. The observed ionization characteristics are consistent with prior reports on boron activation in wide-bandgap semiconductors [1,15,18,31].

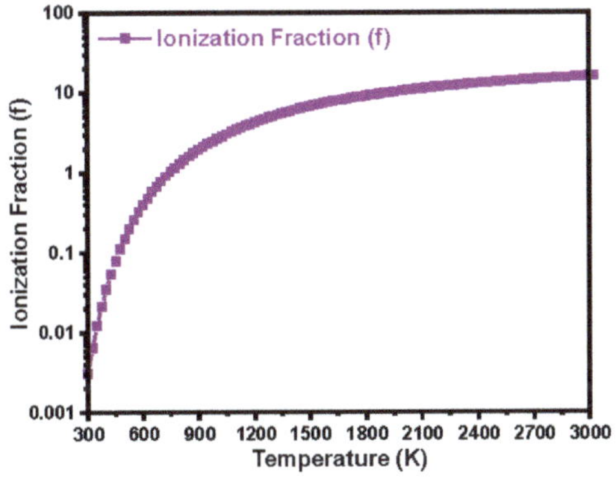


**Figure 7.** Calculated ionization fraction (f) of boron in 4H-SiC as a function of temperature, based on an ionization energy of 298 meV and a degeneracy factor of 4.

*5.4. Current–Voltage Characteristics of the Laser-Fabricated p–n Junction*

The electrical behavior of the laser fabricated p–n junction diode on 4H-SiC substrate at ambient temperature was assessed using current–voltage (I–V) measurements under both forward- and reverse-bias conditions, performed with a 2450 SourceMeter (Keithley Instruments, Inc., Solon, OH, USA). The I–V plot in Figure 8 demonstrates the classical rectifying behavior of a typical diode, with a negligible leakage current under reverse bias and a rapid exponential increase in current under forward bias. This behavior is consistent with the formation of a well-defined p–n junction and confirms the diode's unidirectional conduction property [35,36]. Under forward bias, the current begins to increase significantly beyond ~15 V, marking the turn-on voltage. This relatively high turn-on threshold is attributed to the wide bandgap of 4H-SiC (~3.23 eV) [15], which requires a higher voltage to inject sufficient carriers across the junction. As the forward voltage increases beyond 15–20 V, the current rises steeply, reaching ~69 μA at 35 V.

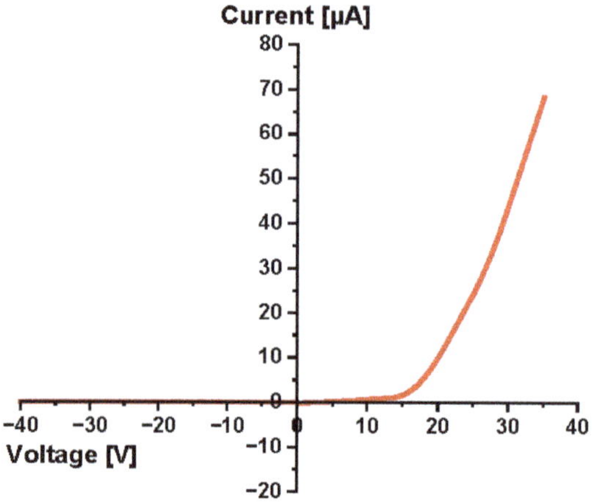


**Figure 8.** I-V characteristics of a laser-fabricated boron-doped p–n junction in n-type 4H-SiC.

In the reverse-bias region, the diode demonstrates excellent blocking capabilities, with a reverse-leakage current of ~0.2 nA across the voltage range up to −40 V. There is no sharp increase in current in the negative voltage region up to −40 V. Hence, no breakdown is observed within this range, suggesting that the diode has a reverse-breakdown voltage well beyond −40 V. The theoretical breakdown voltage $(V_{BR})$ for the laser-fabricated p–n junction in 4H-SiC was calculated using the expression, $V_{BR} = \frac{\varepsilon_s E_C^2}{2qN_D}$, where $\varepsilon_s$ (dielectric permittivity of 4H-SiC) = 9.7 $\varepsilon_0$, $E_C$ (critical electric field for 4H-SiC) = 2.5 × $10^6$ V/cm for moderate doping, q is the electronic charge and $N_D = 10^{16}$ cm$^{-3}$. Substituting these values, the resulting $V_{BR}$ was found to be 1668 V. This value represents the ideal avalanche breakdown condition for a one-sided abrupt ($p^+$–n) junction. This aligns with the expectations for SiC-based diodes, which are known for their high breakdown fields and minimal reverse leakage [35,36]. The built-in potential $(V_{bi})$ was estimated using the standard expression, $V_{bi} = \frac{kT}{q} \ln\left(\frac{N_A \bullet N_D}{n_i^2}\right)$, where $N_A$ and $N_D$ represent the acceptor and donor concentrations, $n_i$ is the intrinsic carrier concentration of 4H-SiC, k is the Boltzmann constant, and T is the room temperature in Kelvin. The $N_A$ (average) was determined to be 5.41 × $10^{17}$ cm$^{-3}$ from the SIMS measurements (see Section 5.3), whereas $n_i$ has a typical value of 8.2 × $10^{-9}$ cm$^{-3}$ at 300 K [37]. Hence, $V_{bi}$ was calculated to be 2.952 V which aligns with the corresponding values reported in [38,39]. This relatively high built-in potential is characteristic of the wide-bandgap semiconductors and supports efficient charge separation across the junction.

The extracted electrical parameters confirm the viability of the laser doping technique for forming functional p–n junctions in SiC substrates. The high turn-on voltage, excellent reverse-blocking behavior, and exponential forward-conduction reinforce the potential of this device for high-voltage and high-temperature power electronics applications.

### *5.5. Determination of Refraction and Attenuation Indices Pre- and Post-Laser Doping*

As illustrated in Figure 9a, boron was selected as the p-type dopant for the 4H-SiC substrate due to its ability to introduce a discrete acceptor energy level, $E_A = 0.29$ eV, above the valence band, corresponding to a photon wavelength of $\lambda = 4.3$ μm within MIDIR spectrum [40,41]. Upon exposure to electromagnetic radiation at this wavelength, valence band electrons are excited into the boron-induced acceptor level, thereby increasing the electron density ($N_e$) and modulating both the refraction index (n) and attenuation index (k) of the doped SiC [1]. This modulation is inherently dependent on the wavelength of the incident electromagnetic radiation and the natural frequency of electrons within the semiconductor [23,42,43], highlighting the potential application of boron-doped 4H-SiC as an MIDIR-responsive material. The optical response before and after doping was evaluated using a PerkinElmer Spectrum-2 FTIR spectrometer (Figure 9a,b). Given that the as-received substrate exhibits negligible transmittance (T~0%) within the MIDIR range, the increase in $N_e$ primarily alters the reflectance (R), through which absorptance (A) can be determined as follows: $A = 100 - R$.

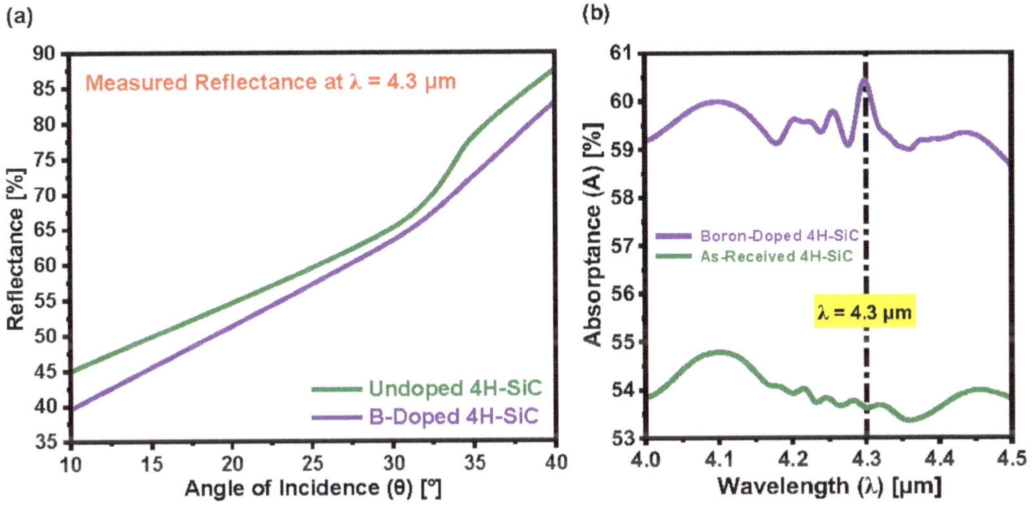


**Figure 9.** (**a**) Measured reflectance (R) of as-received and boron-doped 4H-SiC substrates at $\lambda = 4.3$ μm for angle of incidence (θ), illustrating the influence of boron incorporation on angular reflectance behavior. (**b**) Absorptance (A) of SiC before and after doping, exhibiting a pronounced absorption peak at $\lambda = 4.3$ μm attributed to electronic transitions involving the boron-induced acceptor level.

Figure 9a shows the reflectance as a function of incidence angle, revealing a monotonic increase for both samples. Across all measured angles, the boron-doped SiC region ($R_D$) consistently exhibits lower reflectance compared to the as-received (undoped) SiC region ($R_U$) of the substrate. This reduction in reflectance is indicative of altered optical constants, particularly the refraction and attenuation indices, due to the introduction of free carriers and impurity states resulting from boron incorporation. The modified optical behavior arises from the enhanced interaction between incident radiation and electronic transitions within the doped semiconductor. The angular reflectance profile further confirms that boron incorporation effectively tailors the surface optical response of 4H-SiC, which is critical for optoelectronic applications in the MIDIR range. Figure 9b presents the corresponding absorptance spectra, where the boron-doped sample exhibits a pronounced absorption peak at $\lambda = 4.3$ μm—consistent with transitions from the valence band to the boron-induced acceptor level ($E_A$ ~0.29 eV). At $\lambda = 4.3$ μm, the absorptance increases from 53.62% in the as-received substrate to 60.43% in the doped sample, confirming successful dopant incorporation. Additionally, a secondary absorption feature emerges near $\lambda$ ~4.8 μm (corresponding to ~0.258 eV), which may be attributed to interstitial boron, variation in substitutional doping at Si or C lattice sites, or local lattice strain induced by atomic size mismatch [1]. These mechanisms can introduce additional energy levels within the bandgap, enhancing MIDIR absorption. Moreover, the presence of residual impurities or defects within the substrate, possibly activated during the laser doping process, may contribute to this response [1]. To suppress such extrinsic effects, utilizing a high-purity SiC epitaxial layer as the starting substrate is recommended, as epitaxial layers generally contain fewer impurities and defects compared to bulk-grown substrates [1].

As outlined in Section 4, theoretical absorption models were developed to calculate the refraction and attenuation indices of 4H-SiC substrates, both before and after laser doping. Figure 10a presents the refraction index ($n_2$) and attenuation index ($k_2$) for the as-received (undoped) 4H-SiC substrate at various incident angles. Figure 10b shows the corresponding average values of $n_2$ and $k_2$ across different angles. At $\lambda = 4.3$ μm, the extracted average optical constants are $n_{2,avg} = 2.857$ and $k_{2,avg} = 3.70 \times 10^{-3}$. In the visible and NIR regions, $n_2$ exhibits a decreasing trend with decreasing wavelength due to the normal dispersion, leading to stronger light-bending. Within the MIDIR region, $n_2$ increases due to electronic transitions at shorter wavelengths (~3 μm), resonance effects, phonon interactions at longer wavelengths (~5 μm), and lattice absorption. $k_2$ quantifies material absorption and is predominantly governed by free carrier absorption (FCA) and strong electronic transitions. Electronic transitions occur when electrons move between energy levels in a material due to photon absorption. Here, strong electronic transitions refer to the highly probable and efficient excitations of electrons within the band structure of 4H-SiC, particularly around ~3 μm. At these wavelengths, photon energies are higher, making it easier to excite electrons from the valence band to higher-energy states, including impurity levels introduced by doping. FCA is a phenomenon in which free charge carriers—electrons in the conduction band and holes in the valence band—absorb incident photons, leading to increased optical absorption without requiring interband electronic transitions across the bandgap. Unlike interband absorption, which involves electron excitation from the valence band to the conduction band, FCA occurs within the same band and is strongly dependent on the density and mobility of the free carriers. At ~3 μm, $k_2$ is high due to FCA and the strong electronic transitions. However, as the wavelength increases, the photon energy decreases, reducing the efficiency of these electronic transitions and thereby lowering optical absorption. This complex interplay between electronic transitions and FCA defines the spectral behavior of both $n_2$ and $k_2$, as observed in Figure 10a,b.

Figure 11a illustrates the variation in the refraction index ($n_1$) and attenuation index ($k_1$) for the boron-doped region of the 4H-SiC substrate, extracted using Equation (18) from the absorption models in Section 4. Both $n_1$ and $k_1$ exhibit notable spectral dependence across the MIDIR range, with measurements taken at multiple angles of incidence. The corresponding average values are presented in Figure 11b, where at $\lambda = 4.3$ μm, the extracted values are $n_{1,avg} = 2.485$ and $k_{1,avg} = 12.94 \times 10^{-3}$. The pronounced absorption peak observed in $k_1$, particularly near 4.3 μm, reflects enhanced free carrier absorption (FCA) resulting from boron incorporation. The observed decrease in $n_1$ across the MIDIR spectrum is primarily attributed to the formation of a boron-induced acceptor level above the valence band of 4H-SiC, which significantly modifies the electronic and optical behavior of 4H-SiC. When boron atoms occupy substitutional sites during laser-induced thermal diffusion, they increase the hole concentration at the acceptor level, thereby elevating the free carrier density. This increase in carrier density intensifies FCA, wherein incident photons are absorbed by charge carriers, and the energy is dissipated through phonon emission. The enhanced FCA leads to increased optical damping, thereby reducing the SiC's refraction index by diminishing its capacity to sustain coherent electromagnetic wave propagation. Furthermore, the redistribution of charge carriers associated with doping reduces SiC's electronic polarizability, subsequently decreasing the dielectric constant and contributing to the observed reduction in $n_1$. The inset in Figure 11b illustrates the dependence of the average optical constants on angular frequency ($\omega$), further emphasizing the frequency-sensitive nature of carrier-induced absorption and refraction index suppression. These results collectively underscore the complex interplay between boron-induced electronic structure modifications, dielectric behavior, and optical response in 4H-SiC.

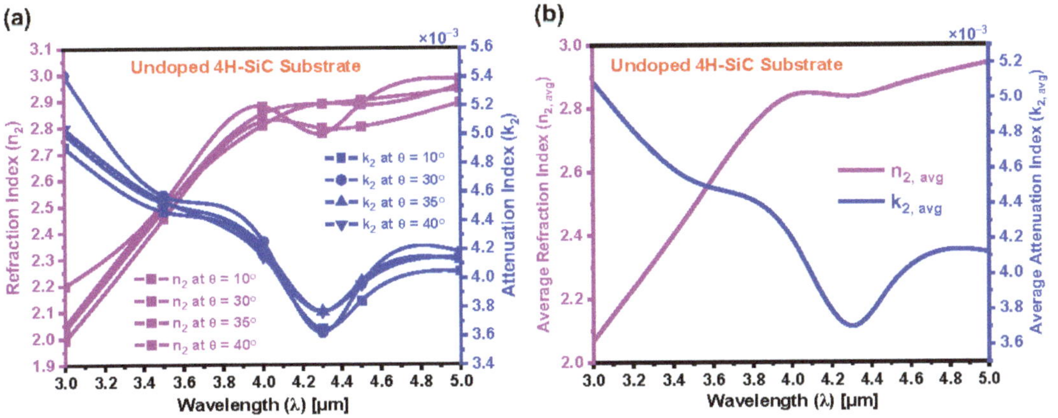


**Figure 10.** (**a**) Refraction index ($n_2$) and attenuation index ($k_2$) of the as-received (undoped) region of the 4H-SiC substrate, measured as a function of wavelength ($\lambda$) at varying angles of incidence. (**b**) Average values of $n_2$ and $k_2$ values over the MIDIR spectral range.

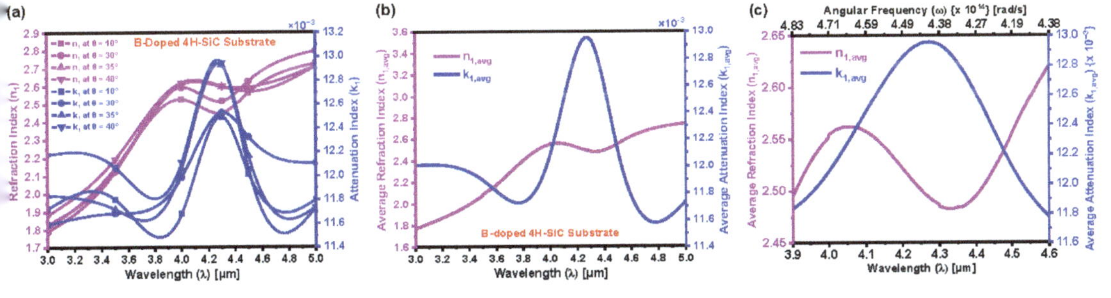


**Figure 11.** (**a**) Refraction index ($n_1$) and attenuation index ($k_1$) of the boron-doped region of the 4H-SiC substrate, measured as a function of wavelength ($\lambda$) at varying angles of incidence. (**b**) Average values of $n_1$ and $k_1$ over the MIDIR range plotted against $\lambda$; (**c**) Corresponding dependence of the average $n_1$ and $k_1$ values on the angular frequency ($\omega$), illustrating the dispersion behavior in the frequency domain.

Figure 12 provides a comprehensive comparison of the n and k values, emphasizing the changes that occur before and after laser doping. This clearly demonstrates how the optical properties are modified as a result of the incorporation of the boron atoms into SiC, offering a visual depiction of the variations in both the real (n) and imaginary (k) components of the complex refraction index. Table 1 provides a quantitative comparison of the refraction index deviations (Δn) and attenuation index deviations (Δk) between the as-received (undoped) and boron-doped regions. These deviations were calculated using both Fresnel's method [21] and the absorption models detailed in Section 4. The Δn and Δk obtained using both methods are minimal, with error margins of $\sim 0.016$ and $\sim 0.46 \times 10^{-3}$, respectively, thereby validating the accuracy of our absorption models.

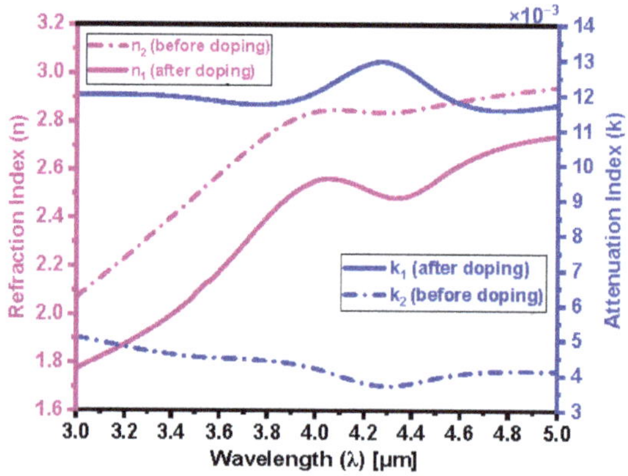


**Figure 12.** Comparison of n and k for the 4H-SiC substrate before and after laser doping. The dashed curves represent the undoped (as-received) region, while the solid curves correspond to the boron-doped region of the substrate.

**Table 1.** Comparison of the refraction index deviations (Δn) and attenuation index deviations (Δk) between the as-received and boron-doped regions of the 4H-SiC substrate.

| | | Fresnel's Method | | | Absorption Models [Current Study] | | |
|---|---|---|---|---|---|---|---|
| **Wavelength (λ)** | **Parameter** | **As-Received** | **Boron-Doped** | **Δ** | **As-Received** | **Boron-Doped** | **Δ** |
| 4.3 μm | n | 2.873 | 2.517 | 0.356 | 2.857 | 2.485 | 0.372 |
| | k | $3.75 \times 10^{-3}$ | $12.53 \times 10^{-3}$ | $8.78 \times 10^{-3}$ | $3.70 \times 10^{-3}$ | $12.94 \times 10^{-3}$ | $9.24 \times 10^{-3}$ |

### *5.6. Determination of Dielectric Constant to Study Anomalous Dispersion in SiC*

The variations in the complex permittivity $\varepsilon_r$, defined as $\varepsilon_{r,real} = n_1^2 - k_1^2$ (real part) and $\varepsilon_{r,img} = 2n_1k_1$ (imaginary part), with respect to angular frequency ($\omega$) are depicted in Figure 13. Under typical conditions, $\varepsilon_r$ tends to increase as $\omega$ increases, a behavior referred to as normal dispersion [23,42]. This trend occurs because, at higher frequencies, the material's ability to polarize in response to the electric field enhances, leading to an increased dielectric constant. However, an intriguing phenomenon arises in the vicinity of the resonance frequency $\omega_0$, where the relationship between $\varepsilon_r$ and $\omega$ reverses, resulting in a decrease in $\varepsilon_r$ as $\omega$ increases. This behavior is classified as anomalous dispersion. Such anomalous dispersion signifies strong interactions between light and the electronic structure of the material, particularly as light approaches the frequency of resonance. These interactions can result in unique optical characteristics that may enhance the performance of photonic devices. As shown in Figure 13, the width of the anomalous dispersion region is $\Gamma = 0.486$ μm. Given that this region is quite narrow, $\omega_0$ can be effectively equated to $\omega_m$, which is the frequency at which $\varepsilon_r$ attains its maximum or minimum value [42]. The difference $(\omega_0 - \omega_m)$ is expressed as $\pm \Gamma/2$, with $|\omega_0 - \omega_m|$ representing the half-width of the anomalous dispersion region, with Γ being the full width [23,42]. Grasping the nature of this anomalous dispersion is crucial, as it can significantly influence the optical properties and potential applications of SiC in a range of photonic devices. For example, the distinctive behavior observed near the resonance frequency can be harnessed in applications like nonlinear optics, waveguiding, and optical sensing, where precise control over light propagation is essential. The complex interplay between electronic transitions and the refraction index in relation to anomalous dispersion provides valuable insights for the development of advanced SiC-based devices in the field of photonics.

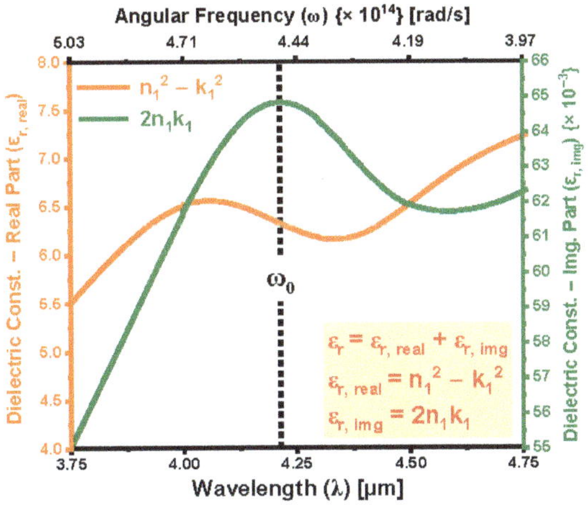


**Figure 13.** Dispersion curves showing the variation in the real and imaginary parts of the dielectric constant ($\varepsilon_r$) as functions of wavelength (λ) and angular frequency (ω).

Boron doping profoundly transforms the electronic band structure and optical behavior of 4H-SiC, playing a pivotal role in advancing its functionality for optoelectronic applications. By introducing extrinsic recombination centers, boron incorporation alters photoluminescence characteristics, with distinct responses governed by dopant concentration and spatial distribution. These modifications reflect substantial changes in the electronic landscape of SiC [40]. Laser doping not only enables controlled boron integration but also facilitates defect engineering within the crystal lattice. The localized thermal energy generated during laser irradiation promotes dynamic annealing, effectively reducing defect densities and enhancing crystalline quality. This dual capability significantly improves both optical and electrical performance. Additionally, the angle of laser incidence emerges as a critical parameter, as it directly influences reflection and attenuation coefficients—factors essential for device efficiency. Variations in the angle of incidence can modulate the optical response, underscoring the importance of precise geometric and optical alignment during device fabrication. Boron doping also impacts carrier concentration and mobility, increasing free carrier density and altering electrical conductivity. These electronic changes are intrinsically linked to modifications in the refraction and attenuation indices, particularly in the infrared region, where pronounced shifts in reflectance occur. Furthermore, the presence of boron introduces impurity-related optical transitions, evident through shifts in photoluminescence peak positions and intensities [40]. A comprehensive understanding of these effects—ranging from dopant-induced band structure alterations to angular dependencies in optical response—is essential for optimizing the optoelectronic performance of 4H-SiC.

## 6. Conclusions

This study presented a detailed investigation into a laser-induced liquid-phase boron-doping method for n-type 4H-SiC, utilizing a pulsed Nd:YAG laser ($\lambda$ = 1064 nm). Through the precise optimization of laser process parameters, the controlled thermal diffusion of boron was achieved without exceeding the peritectic temperature of SiC, thereby maintaining the crystalline integrity of the substrate. FTIR spectrometry confirmed the creation of a boron-induced acceptor energy level at 0.29 eV above the valence band, significantly modulating the optical response of the doped substrate in the MIDIR range. comprehensive optical characterization revealed a substantial reduction in the refraction index from 2.857 (as-received) to 2.485 (doped) and an increase in the attenuation index from $3.70 \times 10^{-3}$ to $12.94 \times 10^{-3}$ at $\lambda$ = 4.3 μm. These results were validated through advanced angle-resolved absorption models, which showed strong agreement with the conventional Fresnel-based methods, confirming the accuracy of the developed models. Electrical characterization of the laser-fabricated p–n junction diode further demonstrated the efficacy of the doping technique. The diode exhibited a high reverse-breakdown voltage of 1668 V and an exceptionally low reverse-leakage current (~0.2 nA), indicating excellent rectifying behavior and suitability for high-voltage applications. A junction depth of 450 nm was measured, with SIMS confirming a peak boron concentration of $1.29 \times 10^{19}$ $cm^{-3}$. Notably, the diffusion coefficient ($D_T = 9.83 \times 10^{-8}$ $cm^2/s$) achieved in this study exceeds the values reported under equilibrium conditions by several orders of magnitude, underscoring the advantages of the non-equilibrium laser doping process in enhancing dopant diffusivity. Additionally, the study identified anomalous dispersion behavior in boron-doped 4H-SiC, reflecting strong light–matter interactions and further enhancing its potential for photonic and optoelectronic MIDIR applications. The ability to precisely tailor both optical and electrical properties through this laser doping technique positions 4H-SiC as a promising material for next-generation devices in power electronics and infrared sensing.

**Supplementary Materials:** The following supporting information can be downloaded at: https://www.mdpi.com/article/10.3390/ma18122758/s1.

**Author Contributions:** G.K., Y.B., C.S., R.K., and A.K. designed and planned the experiments. C.S., R.K., and A.K. were responsible for project planning. G.K. conducted experiments, collected data, and plotted the graphs. Y.B. assisted with doping experiments and characterization. G.K., Y.B., R.K., and A.K. analyzed the results, and co-wrote and reviewed the manuscript. All authors have read and approved the final manuscript.

**Funding:** This research was funded by the Naval Air Warfare Center (NAWC), Aircraft Division, USA under Contract No. N6833519P0493 and the Article Processing Charge was funded by MDPI AG, Switzerland.

**Institutional Review Board Statement:** Not Applicable.

**Informed Consent Statement:** Not Applicable.

**Data Availability Statement:** All data and derivations supporting the findings of this study are provided in the article and its Supplementary Information.

**Acknowledgments:** The authors appreciate the Materials Characterization Facility (MCF) and the Center for Research and Education in Optics and Lasers (CREOL) cleanroom at the University of Central Florida (UCF) and express their gratitude to Eduardo Castillo-Orozco for generously dedicating his time to creating the 3D model of the optical setup for laser doping.

**Conflicts of Interest:** The authors declare no conflicts of interest or competing interests to disclose.

## ferences

Lim, G. An Uncooled Mid-Wave Infrared Detector Based on Optical Response of Laser-Doped Silicon Carbide. Ph.D. Thesis, University of Central Florida, Orlando, FL, USA, 2014.
Nishi, K.; Ikeda, A.; Ikenoue, H.; Asano, T. Phosphorus doping into 4H-SiC by irradiation of excimer laser in phosphoric solution. *Jpn. J. Appl. Phys.* **2013**, *52*, 06GF02. [CrossRef]
Tian, Z.; Salama, I.A.; Quick, N.R.; Kar, A. Effects of different laser sources and doping methods used to dope silicon carbide. *Acta Mater.* **2005**, *53*, 2835–2844. [CrossRef]
Handy, E.M.; Rao, M.V.; Holland, O.W.; Chi, P.H.; Jones, K.A.; Derenge, M.A.; Vispute, R.D.; Venkatesan, T. Al, B, and Ga ion-implantation doping of SiC. *J. Electron. Mater.* **2000**, *29*, 1340–1345. [CrossRef]
Kimoto, T.; Cooper, J.A. Appendix C: Major physical properties of common SiC polytypes. In *Fundamentals of Silicon Carbide Technology: Growth, Characterization, Devices, and Applications*; John Wiley & Sons: Singapore, 2014; pp. 521–524.
Järrendahl, K.; Davis, R.F. Materials properties and characterization of SiC. In *Semiconductors and Semimetals*; Elsevier: Amsterdam, The Netherlands, 1998; Volume 52, pp. 1–20.
Kimoto, T. SiC material properties. In *Wide Bandgap Semiconductor Power Devices*; Woodhead Publishing: Cambridge, UK, 2019; pp. 21–42.
Capan, I. Electrically active defects in 3C, 4H, and 6H silicon carbide polytypes: A review. *Crystals* **2025**, *15*, 255. [CrossRef]
Gordon, L.; Janotti, A.; Van de Walle, C.G. Defects as qubits in *3C*- and *4H*-SiC. *Phys. Rev. B* **2015**, *92*, 045208. [CrossRef]
Koutras, K.N.; Peppas, G.D.; Fetsis, T.T.; Tegopoulos, S.N.; Charalampakos, V.P.; Kyritsis, A.; Yiotis, A.G.; Gonos, I.F.; Pyrgioti, E.C. Dielectric and thermal response of $TiO_2$ and SiC natural ester based nanofluids for use in power transformers. *IEEE Access* **2022**, *10*, 79222–79236. [CrossRef]
Mamun, A.A.; McGarrity, M.; Kim, J.H.; Zhao, F. Silicon carbide-based DNA sensing technologies. *Micromachines* **2023**, *14*, 1557. [CrossRef] [PubMed]
Zhang, Q.; Liu, Y.; Li, H.; Wang, J.; Wang, Y.; Cheng, F.; Han, H.; Zhang, P. A review of SiC sensor applications in high-temperature and radiation extreme environments. *Sensors* **2024**, *24*, 7731. [CrossRef]
Bet, S. Laser Enhanced Doping for Silicon Carbide White Light Emitting Diodes. Ph.D. Thesis, University of Central Florida, Orlando, FL, USA, 2008.
Ikeda, A.; Nishi, K.; Ikenoue, H.; Asano, T. Phosphorus doping of 4H SiC by liquid immersion excimer laser irradiation. *Appl. Phys. Lett.* **2013**, *102*, 052104. [CrossRef]
Kulkarni, G.; Bougdid, Y.; Sugrim, C.J.; Kumar, R.; Kar, A. Laser doping of n-type 4H-SiC with boron using solution precursor for mid-wave infrared optical properties. *J. Laser Appl.* **2024**, *36*, 012016. [CrossRef]
Marui, D.; Ikeda, A.; Nishi, K.; Ikenoue, H.; Asano, T. Aluminum doping of 4H-SiC by irradiation of excimer laser in aluminum chloride solution. *Jpn. J. Appl. Phys.* **2014**, *53*, 06JF03. [CrossRef]
Barkby, J.W.; Moro, F.; Perego, M.; Taglietti, F.; Lidorikis, E.; Kalfagiannis, N.; Koutsogeorgis, D.C.; Fanciulli, M. Fabrication of nitrogen-hyperdoped silicon by high-pressure gas immersion excimer laser doping. *Sci. Rep.* **2024**, *14*, 19640. [CrossRef] [PubMed]
Quick, N.R. Novel techniques in synthesis and processing of advanced materials. In Proceedings of the International Symposium on Novel Techniques in Synthesis and Processing of Advanced Materials, Rosemont, IL, USA, 2–6 October 1994; TMS: Warrendale, PA, USA, 1994.
Quick, N.R.; Matson, R.J. Characterization of a ceramic electrical conductor synthesized by a laser conversion process. In Proceedings of the International Conference on Lasers, Quebec City, QC, Canada, 12–16 December 1994; STS Press: Chicago, IL, USA, 1995.
Incropera, F.P.; Dewitt, D.P.; Bergman, T.L.; Lavine, A.S. *Fundamentals of Heat and Mass Transfer*, 7th ed.; John Wiley: Hoboken, NJ, USA, 1985.
Bougdid, Y.; Chenard, F.; Sugrim, J.; Kumar, R.; Kar, A. $CO_2$ laser-assisted sintering of $TiO_2$ nanoparticles for transparent films. *J. Laser Appl.* **2023**, *35*, 012012. [CrossRef]
Stuck, R.; Fogarassy, E.; Muller, J.C.; Hodeau, M.; Wattiaux, A.; Siffert, P. Laser-induced diffusion by irradiation of silicon dipped into an organic solution of the dopant. *Appl. Phys. Lett.* **1981**, *38*, 715–717. [CrossRef]
Hecht, E. *Optics*; Pearson Education India: Chennai, India, 2012.
Born, M.; Wolf, E. *Principles of Optics: Electromagnetic Theory of Propagation, Interference and Diffraction of Light*; Elsevier: Amsterdam, The Netherlands, 2013.
Howell, J.R.; Mengüç, M.P.; Daun, K.; Siegel, R. *Thermal Radiation Heat Transfer*; CRC Press: Boca Raton, FL, USA, 2020.
Nakashima, S.; Harima, H. Raman investigation of SiC polytypes. *Phys. Status Solidi B* **1997**, *202*, 305–320. [CrossRef]
Zhang, Z.; Li, X.; Wang, L.; Zhang, H.; Shen, L. Effects of doping on the vibrational and electronic properties of 4H-SiC: A Raman and theoretical study. *J. Appl. Phys.* **2020**, *128*, 165703.

28. Ferrari, A.C.; Robertson, J. Interpretation of Raman spectra of disordered and amorphous carbon. *Phys. Rev. B* **2000**, *6*
14095–14107. [CrossRef]
29. Sze, S.M.; Li, Y.; Ng, K.K. *Physics of Semiconductor Devices*, 4th ed.; John Wiley & Sons: Hoboken, NJ, USA, 2021.
30. Roccaforte, F.; Fiorenza, P.; Vivona, M.; Greco, G.; Giannazzo, F. Selective doping in silicon carbide power devices. *Materials* **202**
*14*, 3923. [CrossRef] [PubMed]
31. Troffer, T.; Schadt, M.; Frank, T.; Itoh, H.; Pensl, G.; Heindl, J.; Strunk, H.P.; Maier, M. Doping of SiC by implantation of boron a
aluminum. *Phys. Status Solidi A* **1997**, *162*, 277–291. [CrossRef]
32. Pierret, R.F. *Semiconductor Device Fundamentals*; Addison-Wesley: Reading, MA, USA, 1996.
33. Smith, W.F. *Principles of Materials Science and Engineering*, 3rd ed.; McGraw-Hill Inc.: New York, NY, USA, 1995.
34. Kubiak, A.; Rogowski, J. Boron and aluminum diffusion into 4H–SiC substrates. *Mater. Sci. Eng. B* **2011**, *176*, 297–300. [CrossRe
35. Tian, Z.; Quick, N.R.; Kar, A. Laser endotaxy and PIN diode fabrication of silicon carbide. *MRS Online Proc. Libr.* **2005**, *911*, 10
[CrossRef]
36. Tian, Z.; Quick, N.R.; Kar, A. Laser endotaxy in silicon carbide and PIN diode fabrication. *J. Laser Appl.* **2008**, *20*, 106–1
[CrossRef]
37. Shao, S. 4H-Silicon Carbide p-n Diode for Harsh Environment Sensing Applications. Master's Thesis, University of Californ
Berkeley, CA, USA, 2016.
38. Singh, R. Reliability and performance limitations in SiC power devices. *Microelectron. Reliab.* **2004**, *44*, 559–572. [CrossRef]
39. Lim, J.K. Simulation and Electrical Evaluation of 4H-SiC Junction Field Effect Transistors and Junction Barrier Schottky Diod
with Buried Grids. Ph.D. Thesis, Rensselaer Polytechnic Institute, Troy, NY, USA, 2007.
40. Racka, K.; Avdonin, A.; Sochacki, M.; Tymicki, E.; Grasza, K.; Jakieła, R.; Surma, B.; Dobrowolski, W. Magnetic, optical an
electrical characterization of SiC doped with scandium during the PVT growth. *J. Cryst. Growth* **2015**, *413*, 86–93. [CrossRef]
41. Sugrim, J.C.; Kulkarni, G.; Bougdid, Y.; Heylman, K.; Kumar, R.; Kar, A.; Sundaram, K. Numerical investigation of laser dopi
parameters for semi-insulating 4H-SiC substrate. *J. Laser Appl.* **2024**, *36*, 022013. [CrossRef]
42. Mazumder, J.; Kar, A. *Theory and Application of Laser Chemical Vapor Deposition*; Springer Science & Business Media: New York, N
USA, 2013.
43. Cherroret, N.; Chakravarty, A.; Kar, A. Temperature-dependent refractive index of semiconductors. *J. Mater. Sci.* **2008**, *4*
1795–1801. [CrossRef]

**Disclaimer/Publisher's Note:** The statements, opinions and data contained in all publications are solely those of the individu author(s) and contributor(s) and not of MDPI and/or the editor(s). MDPI and/or the editor(s) disclaim responsibility for any injury people or property resulting from any ideas, methods, instructions or products referred to in the content.

materials

MDPI

icle

# nfluence of Annealing on the Properties of $Fe_{62}Ni_{18}P_{13}C_7$ Alloy

eksandra Małachowska [1,*], Łukasz Szczepański [1], Andrzej Żak [2], Anna Kuś [1], Łukasz Żrodowski [3,4], kasz Maj [5] and Wirginia Pilarczyk [6]

1 Faculty of Mechanical Engineering, Wroclaw University of Science and Technology, Wybrzeże Wyspiańskiego 27, 50-370 Wroclaw, Poland
2 Institute of Advanced Materials, Faculty of Chemistry, Wroclaw University of Science and Technology, Wybrzeże Wyspiańskiego 27, 50-370 Wroclaw, Poland; andrzej.zak@pwr.edu.pl
3 AMAZEMET Sp. z o. o. [Ltd.], Al. Jana Pawła II 27, 00-867 Warsaw, Poland; lukasz.zrodowski@amazemet.com
4 Faculty of Materials Science and Engineering, Warsaw University of Technology, ul. Wołoska 141, 02-507 Warsaw, Poland
5 Institute of Metallurgy and Materials Science, Polish Academy of Sciences, ul. Reymonta 25, 30-059 Krakow, Poland; l.maj@imim.pl
6 Faculty of Mechanical Engineering, Silesian University of Technology, ul. Akademicka 2A, 44-100 Gliwice, Poland; wirginia.pilarczyk@polsl.pl
* Correspondence: aleksandra.malachowska@pwr.edu.pl

**Abstract**

In this study, the influence of annealing on the phase evolution and mechanical properties of the $Fe_{62}Ni_{18}P_{13}C_7$ (at.%) alloy was investigated. Ribbons produced via melt-spinning were annealed at various temperatures, and their structural transformations and hardness were evaluated. The alloy exhibited a narrow supercooled liquid region ($\Delta T_x \approx 22$ °C), confirming its low glass-forming ability (GFA). Primary crystallization began at approximately 380 °C with the formation of α-(Fe,Ni) and $Fe_2NiP$, followed by the emergence of γ-(Fe,Ni) phase at higher temperatures. A significant increase in hardness was observed after annealing up to 415 °C, primarily due to nanocrystallization and phosphide precipitation. Further heating resulted in a hardness plateau, followed by a noticeable decline. Additionally, samples were produced via selective laser melting (SLM). The microstructure of the SLM-processed material revealed extensive cracking and the coexistence of phosphorus-rich regions corresponding to $Fe_2NiP$ and iron-rich regions associated with γ-(Fe,Ni).

**Keywords:** metallic glasses; crystallization; ribbons; annealing; devitrification

## 1. Introduction

Iron-based metallic glasses are extensively studied materials due to their high mechanical strength, good corrosion resistance, excellent soft magnetic properties, and relatively low production cost [1]. Traditionally, they are produced in the form of ribbons. This is related to the simplicity of the fabrication process and the low requirements for glass forming ability (GFA) [1]. Pauly et al. [2] demonstrated that SLM can produce components from an Fe-based metallic glass, taking advantage of the high cooling rates inherent in the process. In practice, however, the production of high-quality samples from iron-based metallic glasses remains a significant challenge [3–5]. The most widely researched problems are cracks, pores, and crystallized regions [6]. The crystallization tendency of metallic glasses with low GFA can be mitigated through the application of optimized scanning strategies [7], while crack formation may be reduced by improving the plasticity of the material [8]. Unfortunately, most Fe-based metallic glasses exhibit negligible plastic deformation at room temperature, which, in combination with the high thermal gradients generated during laser processing, promotes crack initiation and propagation. To address this limitation, Zou et al. [9] proposed the incorporation of ductile copper particles into a FeCrMoCB metallic glass matrix, which effectively suppressed crack formation during SLM processing. Another promising approach involves the in situ formation of ductile phases. For example, $Fe_{77}Mo_5P_9C_{7.5}B_{1.5}$ bulk metallic glass (BMG) reinforced with in situ formed ductile α-Fe dendrites exhibited a plastic strain exceeding 30% and a fracture strength above 3.0 GPa [10]. This strategy has also been successfully demonstrated for SLM-processed Zr/Ti-based metallic glasses [11].

One of the popular Fe-based metallic glass systems is the Fe–P–C, which was later studied for partial substitution of Fe with Ni, primarily with the objective of improving the compressive plasticity of bulk metallic glasses (BMGs) [12]. Fe-Ni-P-C systems have been reported to have exceptional plasticity. For instance, Guo et al. [13] reported an engineering plastic strain exceeding 50% for $Fe_{62}Ni_{18}P_{13}C_7$ alloy under quasi-static uniaxial compression. Similarly, Sarac et al. [14] reported up to 50% plastic strain for a similar $Fe_{50}Ni_{30}P_{13}C_7$ composition under uniaxial compression (unfortunately, no tensile data were available due to sample geometry limitations). The enhanced plasticity was attributed to the presence of nanometer-sized γ-FeNi crystals homogeneously distributed within the soft matrix. It has also been shown that increasing the Ni content in Fe-Ni-P-C systems reduces GFA and promotes nanocrystal formation, which leads to local phosphorus enrichment and its subsequent ejection in the form of clusters [14]. Consequently, the addition of phosphorus decreases the embrittlement temperature [15]. Although the Fe-Ni-P-C systems have relatively low GFA (the reported critical diameters reached a maximum of 2.5 mm [16]), it has been shown that the crystallization tendency of metallic glasses with low GFA can be mitigated through the application of optimized scanning strategies [7]. Additionally, the uniformly nanocrystallized alloy may still possess good magnetic properties [17].

This study investigates the thermal stability and annealing behavior of melt-spun $Fe_{62}Ni_{18}P_{13}C_7$ (at.%) ribbons. While Fe-based bulk metallic glasses and Fe–Ni–P–C systems have been previously investigated [12,14], a systematic evaluation of their microstructure and properties—combined with a feasibility study of selective laser melting (SLM)—has not yet been reported. The ribbons were annealed at various temperatures, and phase evolution, microstructure, and hardness were characterized. The objective was to determine the influence of annealing on mechanical properties, particularly hardness, and to clarify the crystallization mechanisms involved. For comparison, preliminary experiments involving selective laser melting (SLM) of the same alloy were also conducted. The results from SLM processing are presented as supplementary observations, while the primary focus remains on the outcomes of the annealing experiments.

## 2. Materials and Methods

The base alloy ingot, with a nominal composition of $Fe_{62}Ni_{18}P_{13}C_7$ (at.%), was prepared by induction melting. The atomic percentages were first converted to weight percentages. Based on these values, high-purity elemental Fe, Ni, P, and C (≥99.9%) (Onyxmet Tomasz Olszewski, Olsztyn, Poland) were precisely weighed and then melted together in an induction furnace under an argon atmosphere to ensure homogeneity of the alloy. Subsequently, the ribbons were produced using the melt-spinning technique (Melt Spinner SC, Edmund Bühler GmbH, Bodelshausen, Germany), which enables ultrafast cooling of the molten alloy under vacuum conditions [18]. The melt-spinning process was carried out under the following conditions: the roller rotation speed was set to 20 m/s, and the casting was performed at a temperature of 1100 °C. The system was initially evacuated to a vacuum level of $7 \times 10^{-2}$ mBar and then pressurized to a working pressure of 800 mBar. The injection pressure was 400 mBar, with the distance between the crucible and the roller maintained at 0.3 mm. The crucible used had a diameter of 20 mm.

The thickness of the cast ribbons was at the level of 30 μm. The cast ribbons were cut into seven samples, each 100 mm in length. Six samples were annealed at 310, 380, 415, 470, 510, and 570 °C for 10 min in an Lt15/12/C450 furnace (Nabertherm GmbH, Lilienthal, Germany). The annealing is a standard method to study phase evolution [19]. The annealing temperatures in this study were selected based on the characteristic transformation ranges identified by DSC analysis, as well as the temperatures at which microstructural changes were observed during the in situ TEM heating experiments described below. One sample was retained in the as-cast state and served as a reference. Additionally, the Selective Laser Melting (SLM) process was carried out using gas-atomized $Fe_{62}Ni_{18}P_{13}C_7$ powder with a particle size distribution of +20/−63 μm, supplied by NANOVAL GmbH & Co. KG (NANOVAL GmbH & Co. KG, Berlin, Germany). The printing process was carried out on an AISI 316 substrate using a Realizer SLM 50 system (Realizer GmbH, Borchen, Germany) equipped with a 120 W continuous-wave 1064 nm fiber laser. The parameters were selected based on the experience with low GFA metallic glass alloy [7]; as reported, the application of a "random hatch scanning" strategy prevents local overheating and ensures uniform heat distribution throughout the sample.

The fabrication procedure consisted of two stages [20]. In the first stage, referred to as "premelting", an XY scanning strategy was employed with a laser speed of 125 mm/s, laser power of 30 W, and a hatch distance of 120 μm. Successive layers were rotated by 90 degrees to ensure uniform bonding. The second stage, "remelting", was performed at a scanning speed of 3000 mm/s, laser power of 120 W, and a hatch distance of 30 μm. This step was essential for achieving a high amorphous phase fraction and could only be effectively implemented following the premelting stage, as spatter generation would otherwise prevent successful processing. The samples were fabricated as chamfered blocks with dimensions of 15 × 15 × 4 mm, directly attached to the build plate to minimize the risk of print delamination and to enhance heat dissipation. The build plate was made of AISI 304 stainless steel.

The phase composition of the samples was determined using X-ray diffraction (XRD) with a Bruker D8 Advance diffractometer (Bruker AXS, Karlsruhe, Germany) employing Cu-Kα radiation (λ = 1.5406 Å). Data were collected over a 2θ range of 20° to 80° with a step size of 0.02°.

Hardness measurements were performed via nanoindentation using an NHT3 nanoindenter (Anton Paar TriTec SA, Corcelles, Switzerland) equipped with a Berkovich diamond tip. The tests were conducted under a maximum load of 50 mN and a loading rate of 100 mN/min. Hardness values were calculated according to the Oliver–Pharr method. For each sample, ten measurements were performed, and the reported value represents the arithmetic mean.

Differential scanning calorimetry (DSC) analysis was conducted using a DSC 8500 instrument (PerkinElmer Inc., Waltham, MA, USA) with a heating rate of 10 °C/min, up to a maximum temperature of 600 °C, under a nitrogen atmosphere. The relative amorphous phase content ($P$) in the heat-treated ribbons was determined by comparing the exothermic enthalpy of heat-treated ribbons ($\Delta H_{heat\_treated\_ribbon}$) to that of the as-cast ribbon ($\Delta H_{as\_cast\_ribbon}$) and it was calculated using the formula $P = \Delta H_{heat\_treated\_ribbon} / \Delta H_{as\_cast\_ribbon} \times 100\ pct$ [21].

Samples for metallographic analysis were sectioned from the printed specimens and prepared using standard metallographic procedures. The microstructure of the printed samples was examined using a Keyence VHX-6000 (Keyence Corporation, Osaka, Japan) digital microscope.

The samples for transmission electron microscopy (TEM) analysis were initially punched into 3 mm disks using a Disc Punch (Gatan Inc., Pleasanton, CA, USA). Since the as-cast ribbons had a thickness of approximately 30 μm, no mechanical pre-thinning was required. Final thinning was performed by electrolytic polishing with a TenuPol (Struers A/S, Ballerup, Denmark), followed by a short ion polishing step (2 min) using a DuoMill system (Gatan Inc., Pleasanton, CA, USA) to remove surface contamination and residual polishing artifacts. TEM observations of the ribbons were carried out using a Hitachi H-800 (Hitachi High-Tech Corporation, Tokyo, Japan) transmission electron microscope equipped with a custom-designed Hitachi heating holder. For in situ heating experiments, the samples were heated from room temperature to 570 °C at a rate of approximately 10 °C/min. Selected Area Electron Diffraction (SAED) patterns were analyzed using the open-source software CrysTBox v.1.1 [22].

For nanoscale microstructure investigations, a ThermoFisher Themis G2 200 kV (Thermo Fisher Scientific, Waltham, MA, USA) analytical electron microscope (STEM) was used. This microscope was equipped with a Ceta camera (Thermo Fisher Scientific, Waltham, MA, USA) for bright-field imaging, a Fischione high-angle annular dark-field STEM detector, and an energy-dispersive spectroscope (EDS) using a ChemiSTEM (Thermo Fisher Scientific, Waltham, MA, USA) for chemical composition analysis. The samples were prepared using the focused ion beam (FIB) technique, utilizing a ThermoFisher Scios 2 Dual Beam (Thermo Fisher Scientific, Waltham, MA, USA) microscope with a Ga+ ion source and an EasyLift lift-out system. The FIB-prepared lamellae were mounted on a copper grid using a Pt gas injection system and thinned to approximately 100 nm to ensure electron beam transparency in the TEM.

## 3. Results

### *3.1. Phase Transformation*

The X-ray diffractograms of the $Fe_{62}Ni_{18}P_{13}C_7$ annealed ribbons are presented in Figure 1.

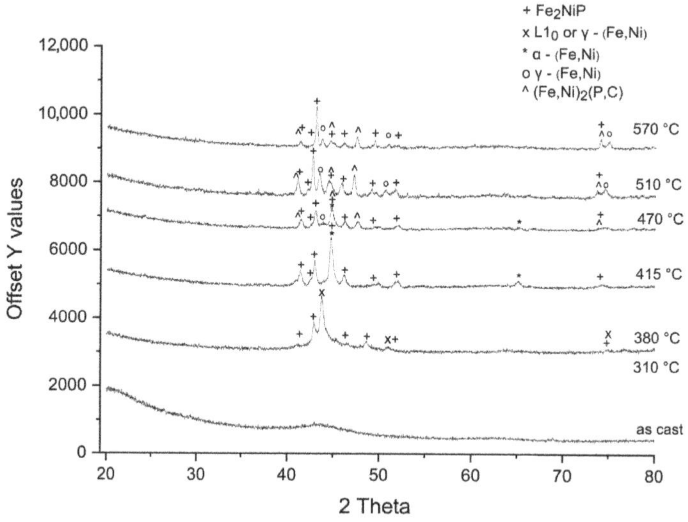


**Figure 1.** X-ray diffraction patterns of $Fe_{62}Ni_{18}P_{13}C_7$ ribbons annealed at given temperatures for 10 min in argon atmosphere, Cu-Kα radiation (λ = 1.5406 Å).

The ribbons remain fully amorphous (no crystalline diffraction peaks) up to an annealing temperature of 310 °C. Upon annealing at 380 °C, new diffraction peaks appear, indicating the nucleation of (Fe,Ni) and the $Fe_2NiP$ phases within the amorphous matrix. Due to overlapping fundamental peaks, it was not possible from XRD alone to unambiguously distinguish whether this intermediate (Fe,Ni) phase has the $L1_0$ or fcc crystal structure. In samples annealed at 415 °C, the intermediate (Fe,Ni) phase is no longer detected. Instead, new diffraction peaks corresponding to the α-(Fe,Ni) phase appear, while those related to the $Fe_2NiP$ phase become more pronounced. At 470 °C, the γ-(Fe,Ni) phase is observed, coexisting with the α-(Fe,Ni) phase. Additionally, alongside the $Fe_2NiP$ phase, the crystallization of another phosphide phase—$(Fe,Ni)_2(P,C)$ is detected. At higher annealing temperatures (510 °C, 570 °C), the γ-(Fe,Ni) phase coexists with phosphides.

In addition to conventional XRD measurements, in situ transmission electron microscopy (TEM) studies were carried out to monitor phase evolution during the devitrification process. The Debye ring patterns recorded during the heating are shown in Figure 2.

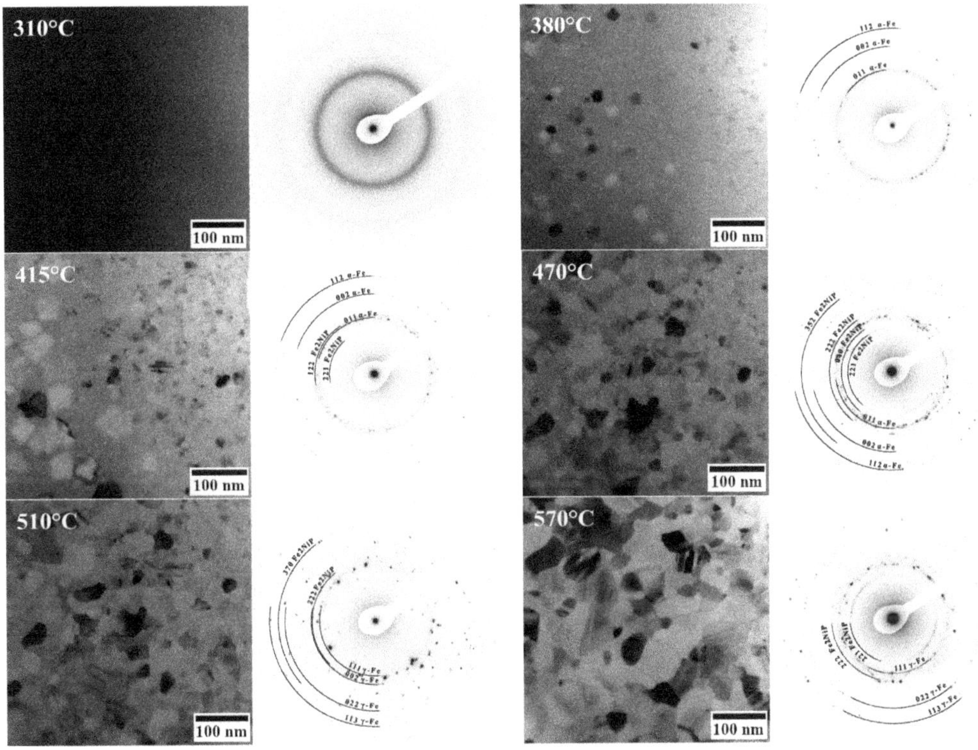


**Figure 2.** Selected area TEM images and corresponding Debye ring patterns recorded during in situ heating at a rate of 10 °C/min of $Fe_{62}Ni_{18}P_{13}C_7$ ribbons, illustrating the phase evolution from the amorphous state to crystalline phases at given temperatures.

At 310 °C, only diffuse halos characteristic of the amorphous phase are observed (no crystalline spots or rings). Upon heating to 380 °C, the first crystallites nucleate within the amorphous matrix. The initial crystalline phase identified in the TEM at ~380 °C is α-(Fe,Ni). At 415 °C, the $Fe_2NiP$ phase appears and coexists with both the α-(Fe,Ni) solid solution and residual amorphous regions (regions between crystallites that have not yet interconnected). Further growth of the $Fe_2NiP$ phase is observed at 470 °C. At 510 °C, the α-(Fe,Ni) phase undergoes transformation into the γ-(Fe,Ni) phase. Twin boundaries, characteristic of austenitic phases, are clearly visible at this stage. At 570 °C, a significant increase in grain size is observed, although no additional new phases are detected. It is noteworthy that there is a discrepancy between the in situ TEM observations and the ex situ XRD results at the 380 °C stage: while XRD indicated the presence of an ordered (Fe,Ni) phase of $L1_0$ or fcc structure, the in situ TEM (with a much shorter effective exposure at 380 °C) showed only α-(Fe,Ni) crystals.

The differential scanning calorimetry (DSC) curve of the investigated alloy is presented in Figure 3.

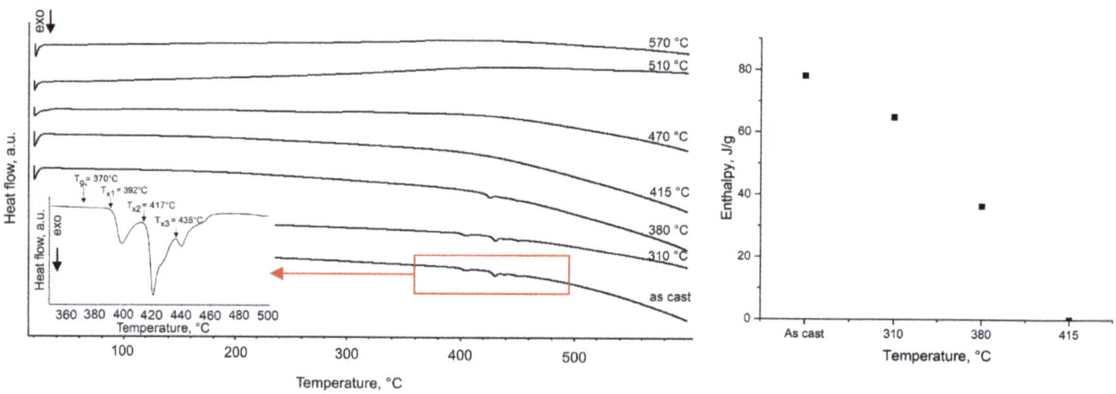


**Figure 3.** DSC curve of the $Fe_{62}Ni_{18}P_{13}C_7$ alloy, indicating the glass transition ($T_g$), the onset of primary crystallization ($T_{x1}$), the second ($T_{x2}$) and third ($T_{x3}$) crystallization events, along with the curves for ribbons annealed at the specified temperatures. Measurements were carried out at a heating rate of 10 °C/min in a nitrogen atmosphere (**left**). Evolution of crystallization enthalpy with annealing temperature (**right**).

The glass transition region is narrow, and the width of the supercooled liquid region $\Delta T_x$ (defined as $\Delta T_x = T_x - T_g$, where $T_g$ is the glass transition temperature and $T_x$ is the onset temperature of crystallization) is approximately 22 °C. Three exothermic peaks are observed, with the primary crystallization onset at $T_{x1} \approx 392$ °C, a second crystallization event at $T_{x2} \approx 417$ °C, and a third, smaller exothermic event at $T_{x3} \approx 435$ °C. These multiple exotherms indicate a multistage devitrification process. Analysis of the crystallization enthalpies reveals a progressive decrease in the amorphous phase fraction as annealing temperature increases (Figure 3). The crystallization exothermic enthalpy of a ribbon annealed at 310 °C corresponds to an estimated ~83% remaining amorphous fraction, which drops to ~47% after annealing at 380 °C. Ribbons annealed at or above 415 °C show essentially no residual amorphous phase by this DSC criterion, consistent with those samples being fully crystalline.

### 3.2. Hardness Evolution

The evolution of hardness with annealing temperature is presented in Figure 4.

The hardness of the as-cast ribbon was approximately 8500 MPa. With increasing annealing temperature up to 415 °C, the hardness increases markedly, by about 50%, reaching roughly 12,500 MPa at 415 °C. In the intermediate temperature range between 415 °C and 510 °C, the hardness remains relatively constant, on the order of ~12,500–13,000 MPa. However, at the highest annealing temperature of 570 °C, a significant drop in hardness is observed, with the hardness decreasing by approximately 1500 MPa compared to the

510 °C condition. Thus, the hardness vs. temperature curve shows an initial rise, a plateau between ~415–510 °C, and then a decline at 570 °C.

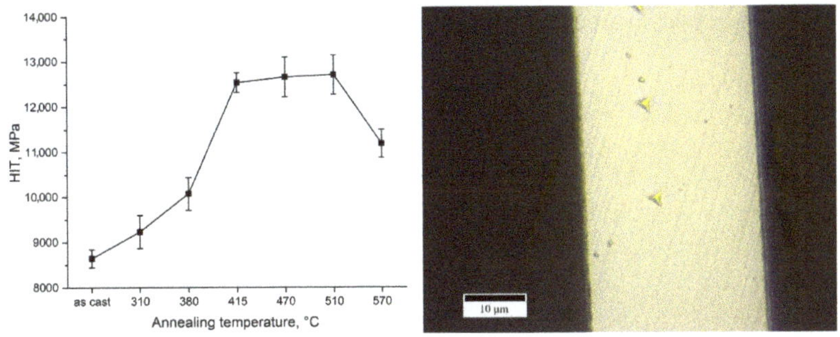


**Figure 4.** Indentation hardness (HIT) obtained from nanoindentation measurements of ribbons in the as-cast state and after annealing (10 min) (**left**) and an example of the obtained indents (**right**).

### 3.3. SLM Printing

The microstructure of the printed sample produced from $Fe_{62}Ni_{18}P_{13}C_7$ powder is shown in Figure 5.

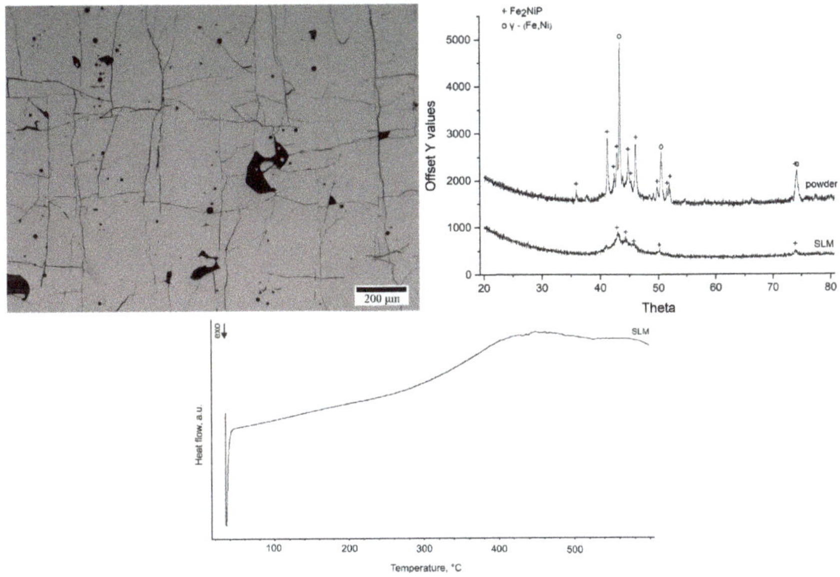


**Figure 5.** Microstructure of SLM printed sample (Keyence VHX-6000 digital microscope) (**left**); diffractogram of powder and printed sample, Cu-Kα radiation (λ = 1.5406 Å) (**right**); DSC curve of the printed sample at a heating rate of 10 °C/min in a nitrogen atmosphere (**bottom**).

It reveals a consistent, irregular square grid of cracks throughout the structure, accompanied by the presence of pores. The phase constitution of the feedstock powder and the SLM-fabricated sample was examined by XRD (Figure 5). The XRD pattern of the as-received powder shows that the alloy powder was predominantly crystalline prior to printing, with strong diffraction peaks corresponding to the γ-(Fe,Ni) and $Fe_2NiP$ phases. This observation is consistent with the low glass-forming ability of $Fe_{62}Ni_{18}P_{13}C_7$ alloy. After SLM processing, the XRD pattern of the printed sample still shows the presence of the $Fe_2NiP$ phase, although its diffraction peaks are less intense compared to the powder. Peaks corresponding to the $\gamma$-(Fe,Ni) phase are absent, suggesting that during laser melting and rapid solidification, part of the $\gamma$-phase present in the powder either partially dissolved or transformed, and $Fe_2NiP$ crystallized from the melt in its place. The DSC curve of the printed sample (Figure 5) indicates that the SLM sample is nearly fully crystallized. Only a very small residual exothermic signal is observed upon heating, which corresponds to an estimated ~1.9% amorphous fraction remaining in the printed material (calculated by comparing the crystallization enthalpy of the printed sample to that of a fully amorphous ribbon).

The measured hardness for the printed sample was 10,200 ± 600 MPa. This hardness value is consistent with a mostly crystalline material that still contains a minor fraction of amorphous phase or is very fine. The presence of a crack grid prevents further mechanical testing of the sample.

Further microscopic analysis of the SLM sample was conducted using STEM-EDS mapping. Elemental mapping in Figure 6 reveals two distinct regions: (i) localized phosphorus-enriched regions with widths up to ~500 nm, although the majority were significantly narrower and (ii) iron-enriched regions with widths up to 600 nm. The phosphorus-enriched zones correspond to the $Fe_2NiP$ phase, and Fe-rich areas might be assigned to the $\gamma$-(Fe,Ni) type solid solution (Figure 7).

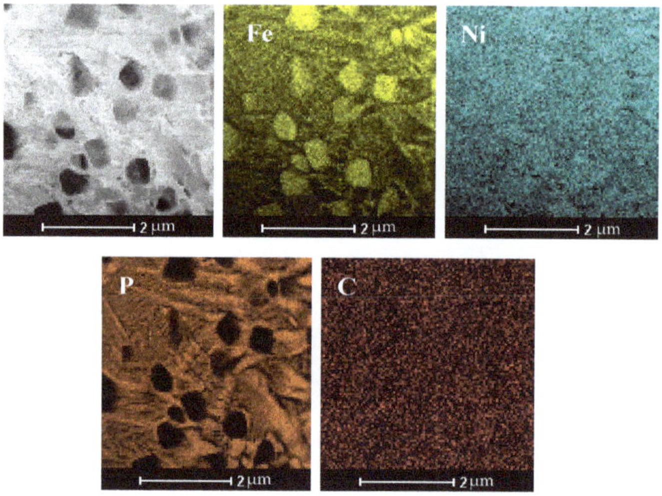


**Figure 6.** STEM microstructure and elemental composition maps in the $Fe_{62}Ni_{18}P_{13}C_7$ SLM printed sample.

From the maps, it is evident that the volume fraction of the $Fe_2NiP$ phase is larger than that of the $\gamma$-(Fe,Ni) phase in these local areas, consistent with the pronounced P segregation. Overall, the microstructure of the SLM-fabricated alloy consists of a brittle $Fe_2NiP$ intermetallic network intermixed with regions of $\gamma$-(Fe,Ni) solid solution. Notably, $\alpha$-(Fe,Ni) was not observed in the as-printed sample; the iron-rich phase present after SLM is in the $\gamma$-(Fe,Ni) form.

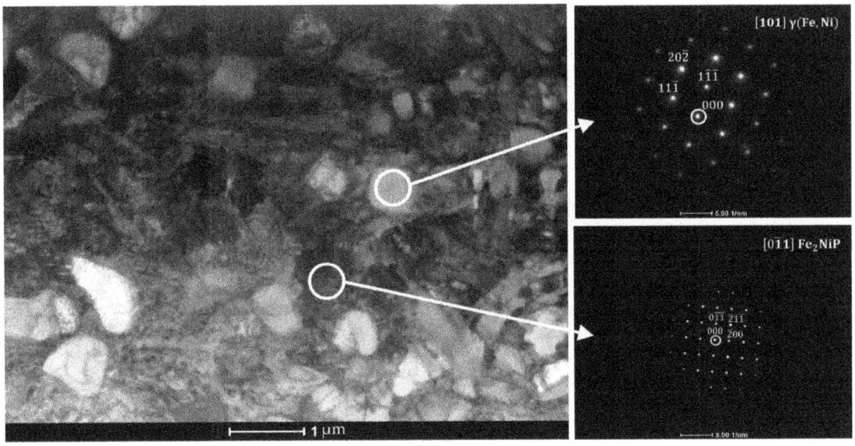


**Figure 7.** Transmission electron microscopy (TEM) image of $Fe_{62}Ni_{18}P_{13}C_7$ SLM-printed sample along with corresponding diffraction patterns identifying the γ-(Fe,Ni) solid solution and $Fe_2NiP$ phosphides.

## 4. Discussion

### *4.1. Phase Transformation*

Upon annealing the $Fe_{62}Ni_{18}P_{13}C_7$ ribbons, the alloy undergoes a multistage crystallization process. The XRD and TEM results together indicate that the first crystalline phase to form is around 380 °C. XRD suggests this phase could be the $L1_0$-FeNi (tetrataenite) or fcc-FeNi (taenite), phase but distinguishing between these ordered structures by conventional XRD is challenging due to overlapping fundamental peaks and the requirement to observe superlattice reflections, which are typically very weak and often fall below the detection limit of conventional XRD measurements [23]. Tetrataenite ($L1_0$-FeNi), if formed, is noteworthy because it is a highly ordered FeNi phase known for its exceptional magnetic properties [24]. It was first discovered in meteorites [25,26]. The $L1_0$-crystal structure belongs to the P4/mmm space group and typically forms near equiatomic compositions (50:50 atomic%) in certain alloy systems. In most cases, the $L1_0$-phase forms via a nucleation-and-growth mechanism from the disordered parent A1 (fcc) phase, below a critical chemical order-disorder temperature [27]. In Fe-Ni alloys, tetrataenite is typically obtained during slow cooling below ~320 °C, as atomic rearrangement from the disordered face-centered cubic taenite phase into a tetragonal distorted face-centered cubic cell requires long-range diffusion [28]. Under natural conditions, such as those found in meteorites, this process can take over $10^4$ years due to the extremely low diffusion rates at these temperatures [28]. In laboratory studies, alternative synthesis routes for tetrataenite have been explored, such as the annealing of metallic glasses [29] or cyclic oxidation-reduction of nickel-coated iron particles [30]. In the case of Fe-Ni-P-C alloys, the crystallization of tetrataenite is believed to be closely related to the presence of phosphorus. Phosphorus significantly enhances the mobility of nickel atoms, particularly at low temperatures in the range of 300–700 °C [31]. The model proposed by Ke et al. [31] predicts a tenfold acceleration of vacancy-mediated solvent diffusion at temperatures below 600 °C when the phosphorus content exceeds 0.1 at.%. Therefore, conditions are favorable for the formation of ordered Fe–Ni phases during annealing of this alloy, despite the kinetic constraints that usually prevent their formation on laboratory timescales.

The appearance of the intermediate (Fe,Ni) phase around 380 °C in the ribbons is accompanied by the onset of nucleation of the $Fe_2NiP$ intermetallic compound, so the

$(Fe,Ni)_3P$ type phase. $Fe_2NiP$ (schreibersite) is a brittle intermetallic compound, with a reported hardness in the range of 800–950 HV [32]. Its formation contributes significantly to the embrittlement of metallic glasses upon crystallization.

By 415 °C, the XRD data show that the initial (Fe,Ni) phase present at 380 °C has disappeared, replaced by a mixture of α-(Fe,Ni) and phosphide phases. This suggests that the intermediate phase was metastable and decomposed upon further heating. At this stage, the predominant phases are α-(Fe,Ni), a ferromagnetic bcc phase, and $Fe_2NiP$. At higher temperatures, the α-(Fe,Ni) phase partially transforms into the γ-(Fe,Ni) phase. The observed sequence of phase evolutions is in agreement with reported phase transformations in Fe–Ni–P alloys [33]. The stabilization of γ-(Fe,Ni) (fcc) over α-(Fe,Ni) (bcc) at elevated temperature is consistent with the Fe–Ni binary phase behavior, where higher Ni content and higher temperatures favor the fcc phase [34]. The crystallization pattern is relevant to the structural relaxation and embrittlement behavior of metallic glasses upon annealing. In particular, $(Fe_{100-x}Ni_x)_{83}B_{17}$ metallic glasses have been shown to exhibit embrittlement only in compositions that crystallize predominantly into the bcc phase [35]. In this study, the $Fe_{62}Ni_{18}P_{13}C_7$ ribbons crystallized primarily into α-(Fe,Ni) at lower annealing temperatures, which would suggest a risk of embrittlement. At higher temperatures, α-(Fe,Ni) transformed to γ-(Fe,Ni). The eventual presence of γ-phase (especially in combination with a fine microstructure) could mitigate some embrittlement, but in practice, the extensive precipitation of $Fe_2NiP$ likely dominates and causes brittleness regardless of α or γ matrix phase.

It is worth noting a discrepancy between the in situ TEM observations and the XRD results for samples annealed at 380 °C. While XRD suggests the formation of an $L1_0$-FeNi or fcc-FeNi type phase, the in situ TEM analysis reveals only the presence of α-(Fe,Ni). This difference is likely attributable to the significantly shorter exposure times during the in situ TEM heating experiments and hence the lack of time for atoms to rearrange.

Thermal parameters from DSC reinforce the conclusion that $Fe_{62}Ni_{18}P_{13}C_7$ has a low glass-forming ability and multistage crystallization pattern. The glass-forming ability (GFA) of an alloy is often characterized parameters such as the critical cooling rate ($R_c$) and critical casting diameter ($D_c$), where $R_c$ represents the minimum cooling rate required to suppress crystallization and achieve a fully amorphous structure, while $D_c$ denotes the maximum sample dimension that allows for the formation of a completely amorphous phase [36]. Additionally, GFA can be estimated using various thermal parameters derived from DSC analysis [37]. One of the most commonly used indicators is the width of the supercooled liquid region ($\Delta T_x$) [37]. Alloys with a small, supercooled liquid region ($\Delta T_x$) require high cooling rates to form an amorphous structure and tend to crystallize easily upon any thermal input. The measured $\Delta T_x$ of 22 °C matches well with literature data. For example, Ma et al. [16] reported the $\Delta T_x$ of 23 °C for a 2.3 mm diameter rod cast from a similar alloy composition, $Fe_{80}Ni_{20}P_{13}C_7$. The fact that the atomized powder was largely crystalline (as shown by XRD) further illustrates the difficulty of retaining an amorphous structure in this composition without extremely high cooling rates or special processing. The DSC curve reveals multiple closely spaced exothermic events, which are indicative of a complex, multistage crystallization process. The first crystallization peak is most likely associated with the formation of the α-(Fe,Ni) phase, followed by a second event corresponding to the nucleation of the $Fe_2NiP$ intermetallic compound. The third exothermic peak is also observed at a higher temperature. Since no new phases are detected by TEM at 470 °C (Figure 2), this final thermal event is likely related to further ordering or the continued crystallization of the $Fe_2NiP$ phase that initially formed around $T_{x2}$.

## 4.2. Hardness Evolution

Annealing has a pronounced influence on the hardness evolution of the $Fe_{62}Ni_{18}P_{13}C_7$ alloy. A ~50% increase in hardness between the as-cast state and samples annealed at 415 °C may be attributed to structural transformations occurring within this temperature range. Specifically, up to 415 °C, nanocrystallization takes place, leading to the formation of α-(Fe,Ni) nanocrystals, as well as the initial precipitation of $Fe_2NiP$ intermetallic, as supported by both XRD and TEM data. The development of these nanometer-scale phases induces classic dispersion strengthening: the presence of hard particles (1000 HV 0.05 for $Fe_3P$ and 1100 HV 0.05 for $Fe_2P$ [38]) and refined α-(Fe,Ni) grains effectively impedes plastic deformation, thereby elevating the overall hardness. This phenomenon aligns with the strengthening behavior observed in other partially crystallized metallic glasses [39–41]. Crystallite size grows from ~20 nm to ~50 nm between 380 °C and 415 °C (Figure 2), enhancing hardness via dispersion strengthening up to an optimal size (~50 nm) [40,41]. Between 415 °C and 510 °C, the alloy retains its elevated hardness, despite continued microstructural evolution. Between 415 °C and 470 °C, the amorphous fraction diminishes further, and the volume fraction of both α-(Fe,Ni) and $Fe_2NiP$ increases. At 470 °C, the alloy becomes fully crystalline, primarily consisting of α-(Fe,Ni) and $Fe_2NiP$. Then, the α-(Fe,Ni) transforms to γ-(Fe,Ni). The α-Fe phase typically exhibits higher hardness than γ-Fe, and its hardness remains more stable with temperature [42]. Two opposing effects appear to balance each other: (i) the transformation of the harder α-phase to the softer γ-(Fe,Ni) phase begins around 470–510 °C, while (ii) the volume fraction of hard $Fe_2NiP$ continues to grow. The retention of significant α-phase content and the presence of additional phosphide precipitates likely offset the inherent softness of the emerging γ-phase, maintaining overall high hardness. At 570 °C, however, a marked drop in hardness is observed due to further reduction of α-phase content and grain coarsening.

## 4.3. SLM Printing

The SLM processing results provide insight into the challenges of additively manufacturing alloys with low GFA, such as $Fe_{62}Ni_{18}P_{13}C_7$. The printed samples suffered from extensive cracking, which is consistent with reports in the literature that Fe-based metallic glass forming alloys tend to crack during laser powder bed fusion [6]. The use of a random hatch remelting strategy was motivated by its previously reported effectiveness in reducing crystallinity in another low-GFA alloy, $Fe_{71}Si_{10}B_{11}C_6Cr_2$ (at.%), as demonstrated by Zrodowski et al. [7]. In the present work, the two-stage scanning approach (premelt followed by remelt) similarly failed to produce a fully amorphous structure in the $Fe_{62}Ni_{18}P_{13}C_7$ alloy. Nonetheless, it appears to have influenced the resulting phase distribution: the printed sample retained a minor amorphous fraction (~1.9%) and exhibited a reduced fraction of the γ-(Fe,Ni) phase compared to the initial powder. Essentially, the laser parameters were sufficient to melt the powder and homogenize it briefly in the melt pool, but as the melt pool cooled and subsequently experienced reheating from nearby scan tracks, the alloy could not avoid crystallization.

The formation of a crack network in the SLM sample can be attributed to several factors. First, the alloy's poor glass-forming ability means it crystallizes during solidification into brittle phases. Second, the rapid cooling and repeated thermal cycling in SLM (as new layers are deposited) generate high internal stresses. The combination of a hard, brittle microstructure and high thermal stresses is conducive to crack initiation and propagation.

The pores observed in the microstructure can be classified into two types according to [43]: (A) small, round pores, typically referred to as hydrogen or metallurgical porosity, and (C) rectangular or triangular pores with sharp edges. The small gas porosity (A) might be minimized through drying of the powder and maintaining a high-purity inert atmosphere. The second type of pores (C) is inherently linked to the thermal stresses exceeding the material's strength, leading to fracture and material detachment during metallographic preparation. This is consistent with the observed high density of cracks observed in the printed sample.

One intriguing aspect of the SLM sample is the presence of the $\gamma$-(Fe,Ni) phase, as opposed to the $\alpha$-(Fe,Ni) that was the primary phase in annealed ribbons. The difference arises from the processing paths: the gas-atomized powder was largely $\gamma$ to begin with (due to its solidification condition), and the SLM process, despite partial remelting, did not convert a $\gamma$-(Fe,Ni) to $\alpha$-(Fe,Ni). Some $\gamma$-(Fe,Ni) phase was retained or re-formed upon solidification of the melt pools. It has been reported in the literature that if the triggering substrate exhibits any crystallographic matching with the phase to be nucleated, it can influence primary phase selection [34], which might also explain the formation of the $\gamma$-(Fe,Ni) phase. An additional factor is the nickel content, at lower Ni contents, the (Fe,Ni) solid solution adopts a body-centered cubic (bcc) structure, whereas at higher Ni contents, it transforms into a cubic close-packed (ccp, $\gamma$-phase) structure in (Fe-Ni)-based metallic glasses [34]. Since the formation of $Fe_2NiP$ consumes more Fe, it may promote the formation of the $\gamma$-(Fe,Ni) phase. The Fe-Ni-P-C systems are susceptible to liquid spinodal decomposition, even during conventional casting [44], which favors the formation of phosphides. Sarac et al. [14] showed that introducing nanometer-sized $\gamma$-(Fe,Ni) precipitates in an Fe–Ni–P–C bulk glass could dramatically increase compressive plasticity, by effectively lowering the shear modulus of the matrix and promoting the formation of numerous shear bands.

In our SLM sample, we have a $\gamma$-phase present along with the phosphides, but this did not translate into any observable improvement in cracking resistance. The crucial difference is the length scale and distribution of phases: in Sarac's work [14], the $\gamma$ precipitates were very fine and uniformly dispersed in an otherwise amorphous matrix, which optimally tuned the mechanical response. In the SLM-processed material, the microstructure is coarse and the brittle phase $Fe_2NiP$ forms a continuous network at the grain boundaries of the $\gamma$-(Fe,Ni). Such a microstructure is extremely prone to cracking because the brittle phase provides easy crack propagation paths, and the $\gamma$-phase islands are too large to arrest or deflect cracks effectively.

Overall, the SLM experiment on $Fe_{62}Ni_{18}P_{13}C_7$ highlights the challenge of processing low-GFA alloys additively. The random remelting strategy, while somewhat effective in reducing crystallinity (as also observed in [7]), was insufficient to achieve an amorphous structure or suppress cracking. The resulting microstructure, characterized by decreased $\gamma$-phase content, increased $Fe_2NiP$, and a minor amorphous fraction, remained predominantly crystalline and brittle. A comparative approach, based on the analysis of annealed ribbons and SLM-processed samples, shows that while controlled annealing enables the formation of a fine nanostructured composite with high hardness, SLM solidification leads to a coarse, phase-separated structure with inferior mechanical resilience. Bridging this microstructural gap is essential for the successful additive manufacturing of such alloys.

## 5. Conclusions

Based on the conducted research, the following conclusions can be drawn:

- The $Fe_{62}Ni_{18}P_{13}C$ alloy exhibits a narrow supercooled liquid region ($\Delta T_x \approx 22$ °C) and readily crystallizes upon heating, indicating low glass-forming ability.
- The annealing process induces multistage crystallization in the alloy. At approximately 380 °C, $\alpha$-(Fe,Ni) (or a metastable (Fe,Ni) phase that subsequently decomposes into $\alpha$-(Fe,Ni) and $Fe_2NiP$) and $Fe_2NiP$ phases are formed. With increasing temperature, $\alpha$-(Fe,Ni) transforms into $\gamma$-(Fe,Ni), accompanied by further phosphide growth.

- Nanocrystallization increases hardness by ~50%, reaching ~12.5 GPa at 415 °C, due to the formation of fine α-(Fe,Ni) grains and $Fe_2NiP$ precipitates. Grain coarsening above 510 °C reduces hardness.
- SLM processing results in a predominantly crystalline, brittle microstructure with only ~2% amorphous content and visible thermal cracking. Hardness of the as-printed sample (~10.2 GPa) reflects its phase composition.

**Author Contributions:** Conceptualization, A.M.; methodology, A.M., Ł.S., A.Ż. and Ł.Ż.; software, Ł.S. and Ł.M.; validation, A.M.; formal analysis, A.M. and Ł.S.; investigation, A.M., Ł.S., A.Ż., Ł.Ż., Ł.M. and W.P.; resources, A.M., A.Ż., Ł.Ż., Ł.M. and W.P.; data curation, A.M., Ł.S., A.K., Ł.M. and W.P.; writing—original draft preparation, A.M.; writing—review and editing, A.M., A.Ż. and W.P.; visualization, A.M. and Ł.S.; supervision, A.M.; project administration, A.M.; funding acquisition, A.M. All authors have read and agreed to the published version of the manuscript.

**Funding:** The presented research results were obtained as part of the research task entitled "*Crystallization in iron-based metallic glasses*" funded by the pro-quality subsidy for the development of the research potential of the Faculty of Mechanical Engineering, Wrocław University of Science and Technology, in the year 2022.

**Institutional Review Board Statement:** Not applicable.

**Informed Consent Statement:** Not applicable.

**Data Availability Statement:** The original contributions presented in the study are included in the article, further inquiries can be directed to the corresponding author.

**Conflicts of Interest:** Author Łukasz Żrodowski was employed by the company AMAZEMET Sp. z o. o. [Ltd.]. The remaining authors declare that the research was conducted in the absence of any commercial or financial relationships that could be construed as a potential conflict of interest.

## eferences


Pei, Z.; Ju, D. Simulation of the Continuous Casting and Cooling Behavior of Metallic Glasses. *Materials* **2017**, *10*, 420. [CrossRef] [PubMed]

Pauly, S.; Löber, L.; Petters, R.; Stoica, M.; Scudino, S.; Kühn, U.; Eckert, J. Processing metallic glasses by selective laser melting. *Mater. Today* **2013**, *16*, 37–41. [CrossRef]

Nong, X.D.; Zhou, X.L.; Ren, Y.X. Fabrication and characterization of Fe-based metallic glasses by Selective Laser Melting. *Opt. Laser Technol.* **2019**, *109*, 20–26. [CrossRef]

Erutin, D.; Popovich, A.; Sufiiarov, V. Comparative Study of Microstructure and Phase Composition of Amorphous-Nanocrystalline Fe-Based Composite Material Produced by Laser Powder Bed Fusion in Argon and Helium Atmosphere. *Materials* **2024**, *17*, 2343. [CrossRef]

Löfstrand, J.; Goetz, I.K.; Marattukalam, J.J.; Hjörvarsson, B.; Nagy, G.; Skårman, B.; Sahlberg, M.; Jönsson, P.E. Stress related magnetic imaging of iron-based metallic glass produced with laser beam powder bed fusion. *Mater. Des.* **2024**, *244*, 113199. [CrossRef]

Zhang, P.; Tan, J.; Tian, Y.; Yan, H.; Yu, Z. Research progress on selective laser melting (SLM) of bulk metallic glasses (BMGs): A review. *Int. J. Adv. Manuf. Technol.* **2022**, *118*, 2017–2057. [CrossRef]

Żrodowski, Ł.; Wysocki, B.; Wróblewski, R.; Krawczyńska, A.; Adamczyk-Cieślak, B.; Zdunek, J.; Błyskun, P.; Ferenc, J.; Leonowicz, M.; Święszkowski, W. New approach to amorphization of alloys with low glass forming ability via selective laser melting. *J. Alloys Compd.* **2019**, *771*, 769–776. [CrossRef]

Zhang, P.; Zhang, C.; Ouyang, D.; Liu, L. Enhancement of plasticity and toughness of 3D printed binary Zr50Cu50 bulk metallic glass composite by deformation-induced martensitic transformation. *Scr. Mater.* **2021**, *192*, 7–12. [CrossRef]

Zou, Y.; Qiu, Z.; Tan, C.; Wu, Y.; Li, K.; Zeng, D. Microstructure and mechanical properties of Fe-based bulk metallic glass composites fabricated by selective laser melting. *J. Non-Cryst. Solids* **2020**, *538*, 120046. [CrossRef]

Guo, S.F.; Liu, L.; Li, N.; Li, Y. Fe-based bulk metallic glass matrix composite with large plasticity. *Scr. Mater.* **2010**, *62*, 329–332. [CrossRef]

Wu, Y.; Wang, H.; Wu, H.H.; Zhang, Z.Y.; Hui, X.D.; Chen, G.L.; Ma, D.; Wang, X.L.; Lu, Z.P. Formation of Cu–Zr–Al bulk metallic glass composites with improved tensile properties. *Acta Mater.* **2011**, *59*, 2928–2936. [CrossRef]

12. Yüce, E.; Sarac, B.; Ketov, S.; Reissner, M.; Eckert, J. Effects of Ni and Co alloying on thermal, magnetic and structural properti of Fe-(Ni,Co)-P-C metallic glass ribbons. *J. Alloys Compd.* **2021**, *872*, 159620. [CrossRef]
13. Guo, S.F.; Qiu, J.L.; Yu, P.; Xie, S.H.; Chen, W. Fe-based bulk metallic glasses: Brittle or ductile? *Appl. Phys. Lett.* **2014**, *105*, 1619 [CrossRef]
14. Sarac, B.; Ivanov, Y.P.; Chuvilin, A.; Schöberl, T.; Stoica, M.; Zhang, Z.; Eckert, J. Origin of large plasticity and multiscale effects iron-based metallic glasses. *Nat. Commun.* **2018**, *9*, 1333. [CrossRef] [PubMed]
15. Walter, J.L.; Luborsky, F.E. The ductile-brittle transition of some amorphous alloys. *Mater. Sci. Eng.* **1978**, *33*, 91–94. [CrossRef
16. Ma, X.; Yang, X.; Li, Q.; Guo, S. Quaternary magnetic FeNiPC bulk metallic glasses with large plasticity. *J. Alloys Compd.* **2013**, *5* 345–350. [CrossRef]
17. Pang, B.; Long, Z.; Long, T.; He, R.; Liu, X.; Pan, M. Accelerated discovery of Fe-based amorphous/nanocrystalline alloy throu explicit expression and interpretable information based on machine learning. *Mater. Des.* **2023**, *231*, 112054. [CrossRef]
18. Budhani, R.C.; Goel, T.C.; Chopra, K.L. Melt-spinning technique for preparation of metallic glasses. *Bull. Mater. Sci.* **1982**, 549–561. [CrossRef]
19. Stoica, M.; Kumar, S.; Roth, S.; Ram, S.; Eckert, J.; Vaughan, G.; Yavari, A.R. Crystallization kinetics and magnetic properties Fe66Nb4B30 bulk metallic glass. *J. Alloys Compd.* **2009**, *483*, 632–637. [CrossRef]
20. Małachowska, A.; Żrodowski, Ł.; Morończyk, B.; Maj, Ł.; Kuś, A.; Lampke, T. Selective Laser Melting of Fe-Based Metallic Glass with Different Degree of Plasticity. *Metall. Mater. Trans. A* **2023**, *54*, 658–670. [CrossRef]
21. Janovszky, D.; Sveda, M.; Sycheva, A.; Kristaly, F.; Zámborszky, F.; Koziel, T.; Bala, P.; Czel, G.; Kaptay, G. Amorphous alloys ar differential scanning calorimetry (DSC). *J. Therm. Anal. Calorim.* **2022**, *147*, 7141–7157. [CrossRef]
22. Kim, J.; Kim, S.; Suh, J.-Y.; Kim, Y.J.; Kim, Y.K.; Choi-Yim, H. Properties of a rare earth free L10-FeNi hard magnet develop through annealing of FeNiPC amorphous ribbons. *Curr. Appl. Phys.* **2019**, *19*, 599–605. [CrossRef]
23. Klinger, M.; Jäger, A. Crystallographic Tool Box (CrysTBox): Automated tools for transmission electron microscopists ar crystallographers. *J. Appl. Crystallogr.* **2015**, *48*, 2012–2018. [CrossRef]
24. Mandal, S.; Debata, M.; Sengupta, P.; Basu, S. $L1_0$FeNi: A promising material for next generation permanent magnets. *Crit. R Solid State Mater. Sci.* **2022**, *48*, 703–725. [CrossRef]
25. Albertsen, J.F.; Knudsen, J.M.; Jensen, G.B. Structure of taenite in two iron meteorites. *Nature* **1978**, *273*, 453–454. [CrossRef]
26. Petersen, J.F.; Aydin, M.; Knudsen, J.M. Mössbauer spectroscopy of an ordered phase (superstructure) of FeNi in an iron meteori *Phys. Lett. A* **1977**, *62*, 192–194. [CrossRef]
27. Zhang, B.; Lelovic, M.; Soffa, W.A. The formation of polytwinned structures in Fe Pt and Fe Pd alloys. *Scr. Metall. Mater.* **1991**, 1577–1582. [CrossRef]
28. Dos Santos, E.; Gattacceca, J.; Rochette, P.; Fillion, G.; Scorzelli, R.B. Kinetics of tetrataenite disordering J. *Magn. Magn. Mat* **2015**, *375*, 234–241. [CrossRef]
29. Sharma, P.; Zhang, Y.; Makino, A. Magnetic Properties of L1 0 FeNi Phase Developed Through Annealing of an Amorpho Alloy. *IEEE Trans. Magn.* **2017**, *53*, 2100910. [CrossRef]
30. Lima, E.; Drago, V. A New Process to Produce Ordered Fe50Ni50 Tetrataenite. *Phys. Status Solidi A* **2001**, *187*, 119–124. [CrossRe
31. Ke, J.-H.; Young, G.A.; Tucker, J.D. Ab initio study of phosphorus effect on vacancy-mediated process in nickel alloys—An insig into $Ni_2Cr$ ordering. *Acta Mater.* **2019**, *172*, 30–43. [CrossRef]
32. Ghosh, S.; Ray, D.; Murty, S. Raghunathpura: An early crystallised IIAB iron with sulphide micro nodules. *Adv. Space Res.* **20** *58*, 1879–1892. [CrossRef]
33. Houghton, O.S.; Loudon, J.C.; Twitchett-Harrison, A.C.; Panagiotopoulos, N.T.; Lampronti, G.I.; Costa, M.B.; Harrison, R.J.; Gre A.L. Reinterpretation of Report of Tetrataenite in Bulk Alloy Castings. *Adv. Sci.* **2025**, *12*, 2408796. [CrossRef] [PubMed]
34. Greer, A.; Walker, I. Primary crystallization in (Fe, Ni)-based metallic glasses. *J. Non-Cryst. Solids* **2003**, *317*, 78–84. [CrossRef]
35. Lewis, B.G.; Davies, H.A.; Ward, K.D. The crystallisation and associated changes in ductility of some Fe-Ni-B glassy alloys. *S Metall.* **1979**, *13*, 313–317. [CrossRef]
36. Li, X.; Kou, S.; Li, C.; Zhao, Y.; Ding, Y. A criterion of glass-forming ability and stability derived from pseudo-four characterist temperatures. *Intermetallics* **2021**, *134*, 107201. [CrossRef]
37. Suryanarayana, C.; Inoue, A. Iron-based bulk metallic glasses. *Int. Mater. Rev.* **2013**, *58*, 131–166. [CrossRef]
38. Nowacki, J. Phosphorus in iron alloys surface engineering. *J. Achiev. Mater. Manuf. Eng.* **2007**, *24*, 57–67.
39. Minnert, C.; Kuhnt, M.; Bruns, S.; Marshal, A.; Pradeep, K.G.; Marsilius, M.; Bruder, E.; Durst, K. Study on the embrittlement flash annealed Fe85.2B9.5P4Cu0.8Si0.5 metallic glass ribbons. *Mater. Des.* **2018**, *156*, 252–261. [CrossRef]
40. Han, J.; Wang, C.; Kou, S.; Liu, X. Thermal stability, crystallization behavior, Vickers hardness and magnetic properties Fe–Co–Ni–Cr–Mo–C–B–Y bulk metallic glasses. *Trans. Nonferrous Met. Soc. China* **2013**, *23*, 148–155. [CrossRef]
41. Pang, L.L.; Inoue, A.; Zanaeva, E.N.; Wang, F.; Bazlov, A.I.; Han, Y.; Kong, F.L.; Zhu, S.L.; Shull, R.B. Nanocrystallization, goo soft magnetic properties and ultrahigh mechanical strength for Fe82-85B13-16Si1Cu1 amorphous alloys. *J. Alloys Compd.* **20** *785*, 25–37. [CrossRef]

Torres, H.; Varga, M.; Ripoll, M.R. High temperature hardness of steels and iron-based alloys. *Mater. Sci. Eng. A* **2016**, *671*, 170–181. [CrossRef]
Kuś, A.; Rajťúková, V.; Pilarczyk, W.; Hudák, R.; Mehner, T.; Maj, Ł.; Lampke, T.; Małachowska, A. First attempt to print Co-based alloys with high glass forming ability by selective laser melting. *J. Alloys Compd.* **2024**, *995*, 174680. [CrossRef]
Yao, K.F.; Zhang, C.Q. Fe-based bulk metallic glass with high plasticity. *Appl. Phys. Lett.* **2007**, *90*, 61901. [CrossRef]

materials

MDPI

*Article*

# Geometric Accuracy and Mechanical Property Enhancement of Fe-Based Alloy Layers in Wide-Beam Laser Direct Energy Deposition

**Bin Hu [1,*], Junhua Wang [2], Junfei Xu [3], Qingyang Wang [1] and Li Zhang [1]**

[1] College of Physics and Electronic Engineering, Xinyang Normal University, Xinyang 464000, China; zhangli1272@163.com (L.Z.)
[2] School of Mechatronics Engineering, Henan University of Science and Technology, Luoyang 471003, China; wangjh@haust.edu.cn
[3] State Key Laboratory of Mechanical Transmission for Advanced Equipment, College of Mechanical and Vehicle Engineering, Chongqing University, Chongqing 400044, China; mecha_xjf@163.com
* Correspondence: hubin.xystar@xynu.edu.cn

**Abstract**

Laser direct energy deposition (LDED) has been widely employed in surface modification and remanufacturing. Achieving high-precision geometries and superior mechanical properties in cladding layers remains a persistent research focus. In this study, an Fe-based alloy was deposited on an AISI 1045 substrate via a wide-beam laser cladding system. Single-track multi-layer samples were prepared with varying z-increment ($Z_d$), interlayer dwell time ($T_I$) and laser scanning speed (V) values. The geometry, microstructure, microhardness and wear resistance of the samples were analyzed. Experimental results showed that an estimated $Z_d$ can ensure a constant standoff distance of the laser head and resulting geometric accuracy improvement. Planar grains form at the layer–substrate bonding interface and transition to columnar grains adjacently, while dendrites and equiaxed grains are distributed in the middle and top regions of the layer. The coating layer exhibits much better wear resistance and friction properties than the substrate. The cooling rate can be substantially increased by either raising V or prolonging $T_I$, resulting in refined grain structures and enhanced microhardness. Real-time monitoring and controlling the mean cooling rate have been demonstrated to be effective strategies for enhancing cladding layer performance.

**Keywords:** direct energy deposition; cooling rate; Fe-based alloy; z-increment; microhardness

## 1. Introduction

Large-scale mechanical components like rolls, gears and other moving parts are typically subjected to sustained compressive forces and frictional wear. These components exhibit significant susceptibility to surface degradation, ultimately compromising the service life of equipment [1]. Therefore, surface modification technologies have been widely utilized to promote wear resistance and fatigue resistance properties. Laser direct energy deposition (LDED), recognized as an advanced coating and remanufacturing technology, has attracted great attention in recent years. Owing to high-quality energy sources, LDED demonstrates significant advantages in terms of minimal heat-affected zones (HAZs), low dilution rates, strong metallurgical bonding and exceptional material processing flexibility [2–4]. Near-net-shaped components with complex geometries can be easily fabricated at low cost and with high efficiency by layer-wise stacking and adaptable deposition orientation. At present, LDED has been increasingly employed in multiple industries, such as aerospace, mold manufacturing, petrochemicals and medical health [5–7].

LDED melts powder or wire materials on a substrate, and compared to wire-fed LDED, powder-based LDED demonstrates superior capabilities in terms of material composition control and dimensional precision [8]. Fe-based alloy powders are typically enhanced by adding alloying elements such as chromium (Cr), nickel (Ni), molybdenum (Mo), boron (B) and silicon (Si). The addition of B and Si effectively lowers an alloy's melting point, while simultaneously forming an oxidation-resistant protective layer on the substrate, and the incorporation of Cr and Mo can notably promote coating hardness and wear resistance [9–13]. The primary advantage of Fe-based alloys lies in their mechanical properties that are comparable to those of nickel- and cobalt-based alloys, while offering abundant material availability and lower costs. AISI 1045 steel is a widely used material for industrial components and equipment; traditional heat treatment processes are usually employed to promote its mechanical performance. In order to obtain better surface quality, depositing Fe-based alloy powder on AISI 1045 substrates via LDED has become a predominant method [14].

To promote the geometric accuracy of LDED, lots of researchers have focused on parameter optimization and process control [15–17]. For multi-layer cladding, the standoff distance of the laser head governs the locations of laser and powder convergence points [18,19]. Mazzarisi et al. [20] reported that cladding height was reduced due to unbefitting standoff distances, and the maximum powder catchment efficiency was attained when powder focalized in the melt pool center. The same findings were reported by Zhu et al. [21] and Yu et al. [22] during thin-wall part fabrication: the laser became defocused and the cladding process deteriorated when the layer growth failed to match the z-increment. Therefore, maintaining a constant standoff distance is beneficial for improving geometric accuracy and material utilization efficiency, and determining a precise z-increment is particularly crucial for muti-layer LDED. Li et al. [23] utilized the initial layer height as the z-increment within an open-loop control system, achieving favorable geometric precision. The layer height exhibits a strong correlation with process parameters. Therefore, the determination of the z-increment must account for potential variations in these parameters. Regression analysis provides an effective methodology for establishing the linear correlation between process parameters and the z-increment [19]. However, this experimental determination approach exhibits limitations in terms of cost-effectiveness and generalizability [24].

Cladding layer properties can be improved via material design [25,26], while substrate preheating and active cooling constitute another widely adopted approach [27,28]. Reindl et al. [28] reported that the whole cladding layer can gain consistent hardness by adopting a substrate preheating method. By comparing air-cooled and water-cooled cladding, Zhang et al. [29] demonstrated significantly enhanced performance in water-cooled cladding layers, with refined grain structures and substantially increased hardness. In addition, interlayer peening can also achieve similar cooling rate enhancement [30]. However, such methods introduce assistive equipment that increases system complexity and uncontrollability. In contrast, process parameter modulation offers a technically simpler approach to optimize cladding layer performance.

Extending the interlayer dwell time has been widely adopted for mechanical property promotion via LDED [31–33]. Foster et al. [31] reported that extending the interlayer dwell time to 40 s refined the microstructure in both Inconel 625 and Ti-6Al-4V, with the latter exhibiting yield and tensile properties comparable to those of wrought plates. With an interlayer dwell time of 360 s, the elongation and hardness of LDED 316L thin-wall parts can be increased by 47% and 13%, respectively [33]. While interlayer dwell time adjustment offers operational simplicity, it concurrently decreases deposition efficiency. Some researchers have explored other approaches to change the real-time cooling rate [34,35]. Farshidianfar et al. [27,34] carried out studies on the relationship between the laser scanning speed and cooling rate. They maintained the desired high cooling rate by implementing closed-loop scanning speed control during deposition, offering a novel approach for tailoring 316L microstructures in DED processes. Nair et al. [35] employed a comparable control strategy in IN718 DED, regulating both the scanning speed and powder feeding rate to maintain cooling rates above threshold values, thereby enhancing the resultant mechanical properties.

In summary, the z-increment and cooling rate critically influence the geometric accuracy, microstructure and mechanical properties of LDED. Cooling rate regulation has attracted significant attention for fabricating high-performance materials. However, it is not very frequently reported in Fe-based alloy LDED. In this study, Fe-based alloy multilayer cladding experiments were conducted with varying z-increments, scanning speeds and interlayer dwell times. The z-increment values were determined through simulation and analytical modeling, with their impact on standoff distance and geometric accuracy subsequently analyzed. Cooling rate enhancement was achieved through regulation of scanning speed and interlayer dwell time. The cross-sectional samples were observed to compare microstructure, hardness and friction performance.

## 2. Materials and Methods

### 2.1. Materials

In this study, Fe-based alloy powder was employed as the cladding material. The powder was prepared by the gas atomization method, with a 0.859 spherical degree, exhibiting a good flowability. The chemical composition of the powder is detailed in Table 1. The particle size distribution and powder morphology characteristics are listed in Table 2.

**Table 1.** Chemical composition of Fe-based alloy powder (wt%).

| C | Cr | Ni | Fe | P | Mo | B | S |
|---|---|---|---|---|---|---|---|
| 0.173 | 17.78 | 2.68 | Bal. | 0.013 | 0.90 | 0.86 | 0.004 |

**Table 2.** Particle size distribution and morphology characteristics of Fe-based alloy powder.

| Size Distribution (μm) | | | | Flow Rate (s/50 g) | Degree of Sphericity |
|---|---|---|---|---|---|
| D10 | D50 | D90 | D97 | | |
| 67.30 | 109.1 | 177.8 | 219.8 | 17.69 | 0.859 |

The substrate material was AISI 1045, which is widely used as the main body of most industrial machines. The substrate plate size was 100 × 100 × 10 mm$^3$. To improve the laser absorptivity, the substrate was polished and cleaned with alcohol and acetone before the cladding experiments. The chemical composition of the substrate material is given in Table 3.

**Table 3.** Chemical composition of AISI 1045 (wt%).

| C | Cr | Ni | Fe | Mn | Si |
|---|---|---|---|---|---|
| 0.42 | 0.2 | 0.255 | Bal. | 0.65 | 0.3 |

## 2.2. Experimental Setup

The experiment was carried out via a wide-beam laser cladding system (Figure 1). The wide laser beam was generated by a 3000 W fiber laser with a wavelength of 1080 nm, and the power stability was maintained at ±2%. The laser spot size was $6 \times 2$ mm$^2$ at the focus point. The laser head was driven by a 6-axis Kuka robot arm as the end operator, and the repeat positioning accuracy was ±0.04 mm. Cooling water was pumped from a chiller to protect the laser and the laser head from overheating. The cladding material was transported by a powder feeder, whose disk rotated at 0–10 r/min. Also, 99.99% argon gas was adopted as a powder carrier gas and shielding gas. The cladding process was monitored by an infrared camera, which was installed at 60 degrees from the horizontal plane.

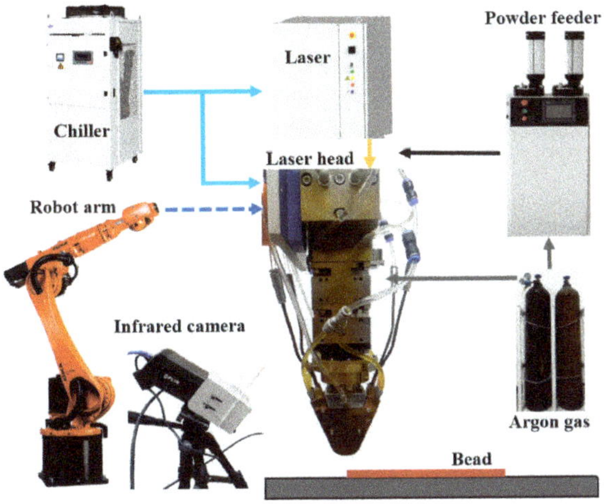


**Figure 1.** Schematic of wide-beam laser cladding system.

## 2.3. Methods

### 2.3.1. Experimental Approach

Single-track multi-layer cladding experiments were carried out to investigate the influences of process parameters on the shape and mechanical properties of the cladding layer. The scanning speed (V), axial lifting amount ($Z_d$) and interlayer dwell time ($T_I$) were chosen as parameters of interest, and three cladding samples were prepared by varying the selected parameters. The laser power (P) was 2000 W, and the powder feeding rate (F) was 22.5 g/min. The powder carrier gas flow rate ($Q_P$) and the shielding gas flow rate ($Q_S$) remained unchanged and were 10 L/min and 12 L/min, respectively. The laser head was 18 mm away from the substrate during the first layer deposition (i.e., initial standoff distance ($D_S$)). The axial lifting amount was equal to the height of a single layer, which was predicted according to our previous research [24,36]. The scanning speed (V) and the interlayer dwell time ($T_I$) were changed to obtain different cooling rates. The process parameters of each sample are listed in Table 4.

**Table 4.** Process parameters of the three samples.

| Sample | P (W) | V (mm/s) | F (g/min) | $Z_d$ (mm) | $T_I$ (s) | $Q_P$ (L/min) | $Q_S$ (L/min) | $D_S$ (mm) |
|---|---|---|---|---|---|---|---|---|
| 1# | 2000 | 7 | 22.5 | 0.45 | 5 | 10 | 12 | 18 |
| 2# | 2000 | 12 | 22.5 | 0.45 | 5 | 10 | 12 | 18 |
| 3# | 2000 | 12 | 22.5 | 0.4 | 50 | 10 | 12 | 18 |

Each sample was 80 mm long and had 5 layers. When one layer was finished, the laser and powder feeder were shut off immediately; then, the laser head rose up according to the preset $Z_d$ and returned back to the track starting point synchronously. The next layer started after waiting for time $T_I$, and the shielding gas was kept flowing during the whole process.

#### 2.3.2. Mechanical Property Analysis

The sample track was cut transversely via a wire cutting machine; a cross section was ground, polished and etched for metallographic inspection. The microstructure was analyzed by an electron microscope. The microhardness was tested along the vertical direction of the cross section, and the sampling points were from the cladding layer surface to the substrate. To evaluate the wear resistance of the cladding layer, friction-wear tests were conducted on the cladding layer and the substrate. The surfaces were ground smooth, without scratches. Silicon nitride balls were used as the friction pair, a pressure of 1000 g was applied and the grinding time was 30 min. The friction coefficient was recorded during the test, and the worn surface was scanned by a three-dimensional topography profilometer.

#### 2.3.3. Cooling Rate Monitoring

Thermocouples, pyrometers and infrared cameras are usually employed to monitor the thermal history of laser cladding. Infrared cameras can implement non-contact measurement online for the deposition process. There is a difficulty in determining the infrared emissivity due to the variation in the surroundings and the surface emissivity of the melt pool. To simplify the calculation, the infrared emissivity was generally considered as a constant value in online monitoring [34]. Thus, temperature measurement deviations were inevitably introduced, and it was necessary to validate the reliability of the temperature measurement. Infrared images can measure the melt pool size with considerable precision based on temperature gradients, which has been validated via combined experimental and simulation approaches in our prior work [24]. Therefore, despite the presence of emissivity errors, infrared images can accurately measure the cooling rate on melt pool surfaces. In this work, thermal video streams were recorded by an infrared camera, and the cooling rate was calculated according to the method proposed by Farshidianfar et al. [34,37]. The images were 800 $\times$ 600 pixels, where each pixel corresponded to the real-time radiation temperature. The picture was converted to an 8-bit grayscale image, and the temperature of the pixels was determined based on the gray value between 0 and 255.

Figure 2a shows the two-dimensional distribution of the pixel array, and X and Y are the numbers of rows and columns, corresponding to 800 $\times$ 600 pixels. The location of each pixel can be expressed as M $(i, j)$, where $i$ is the column vector and $j$ is the row vector. The main region of interest (ROI) is the melt pool, whose occupied pixels should be increased as much as possible in the infrared image. A practical approach is to reduce the view field of the camera; in the experiment, the camera was installed approximately 30 cm away from the deposition position.

The pixel with the maximum gray value has the highest temperature, which can be considered as the center of the melt pool. The method of measuring the cooling rate online via an infrared image is illustrated in Figure 2b. Two frames of images were captured at different times ($t_0$ and $t_1$). The points O and M are the melt pool center and a fixed location on the cladding track, respectively. At time $t_0$, the first image was recorded by the infrared camera, where O is the melt pool center, M represents another point on the melt pool whose location is marked as M $(i_0, j_0)$, and the temperature of M is $T_0$ at this moment. At $t_1$ the next image was captured, and the melt pool had moved a certain distance relative to the previous image. The melt pool center (O) was replaced by O′ in the new image.

The position vector of M was unchanged, and the location was marked as M ($i_1$, $j_1$). The temperature of M dropped to $T_1$ due to cooling down. The cooling rate of any location (M) on the melt pool can be expressed as follows [34]:

$$C_M(t) = \left| \frac{T_1 - T_0}{t_1 - t_0} \right|, \quad (1)$$

It is worth noting that the infrared image only offers the radiation temperature of the melt pool surface, which deviates from the actual temperature. However, the measurement error has no influence on the calculation of the cooling rate, according to Equation (1). Furthermore, the calculated cooling rate is only for the locations on the melt pool surface; the inner cooling rate of the melt pool is unobtainable.

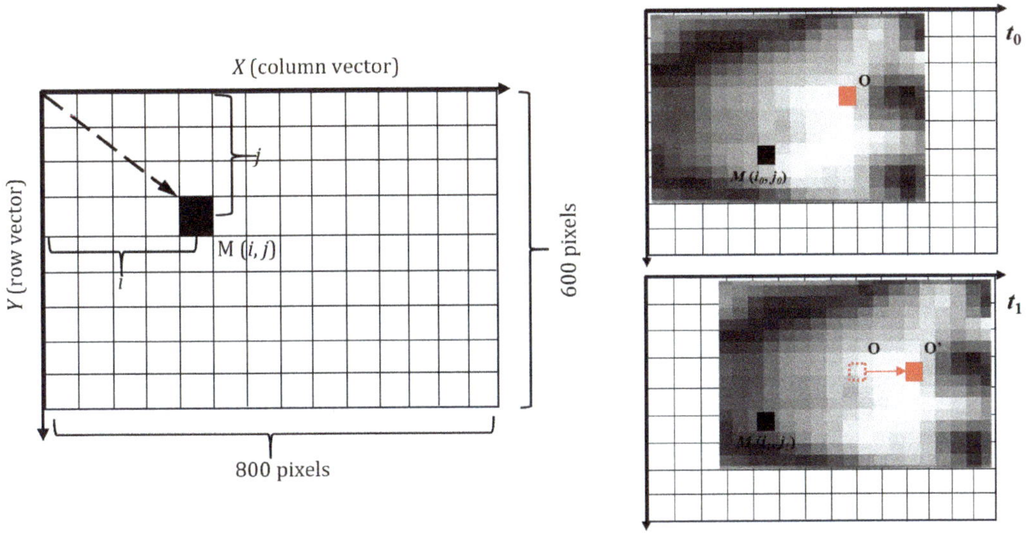


**Figure 2.** Method of cooling rate online monitoring via infrared images. (**a**) Pixel array distribution in an infrared image of A615. (**b**) Cooling rate measurement of a fixed location on the cladding track.

## 3. Results and Discussion

### *3.1. Analysis of Morphological Characteristics*

The macroscopic features of the samples are displayed in Figure 3. No obvious voids or defects were observed on the surface of any of the samples. A few ablative powders adhered to the surface of the cladding layer, and many agglomerate powders could be found on both sides of the cladding track, where poor metallurgical bonding with the substrate was exhibited. The cladding layers on the top became narrower than those on the bottom, as the laser spot was wider than the cladding track; therefore, the powder remaining near the track was melted inadequately by the edge of the laser beam at a low intensity. As can be seen in Figure 3a, some black particles were distributed on the bead surface; this was due to the ablation and oxidation of the powder. In Figure 3b, the agglomerate powders on the track edge of Sample 3 are more abundant than those on Sample 2. This is because the residual powder is more difficult to melt sufficiently in the case of the long cooling time for the substrate (50 s).

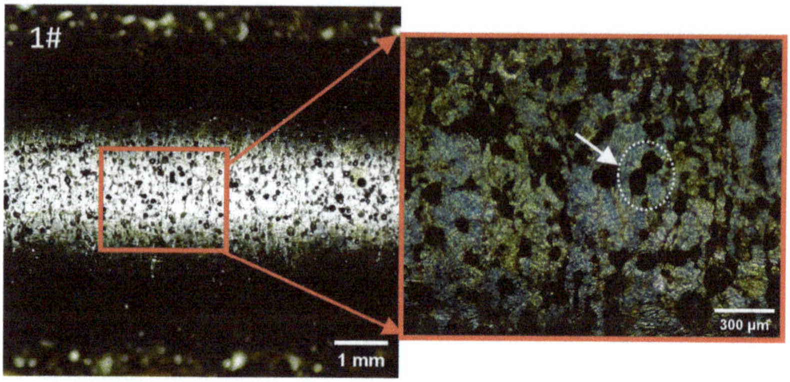


(**a**)

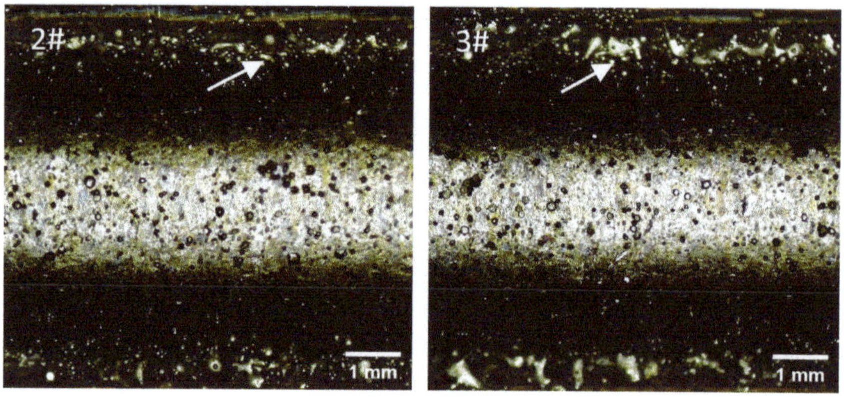


(**b**)

**Figure 3.** Macroscopical observation of the samples. (**a**) Powder ablation and oxidation (marked by dotted circle and white arrow) on the surface of Sample 1. (**b**) Agglomerate powders (marked by white arrow) on the track edges of Samples 2 and 3.

### *3.2. Cross-Sectional Analysis of the Cladding Layer*

#### 3.2.1. Layer-Wise Thickness and Axial Lifting Amount

The cladding layer was sectioned transversely for further characterization. The resulting cross-sectional morphology of each sample is presented in Figure 4. The examination revealed no detectable defects (e.g., voids or cracks).

Due to obtaining a higher laser specific energy, Sample 1 exhibited a more pronounced dilution zone compared to the other samples, indicating good metallurgical bonding with the substrate. Distinct fusion lines could be found across all the samples. The fusion line was generated due to reheating and remelting of the pre-built layer, corresponding to a narrow band with more coarse grains than the surrounding area [38]; fusion lines have been commonly utilized as demarcation boundaries between adjacent layers in prior research [39]. Figure 4a illustrates the five layers marked by dash lines; the top layer demonstrates a significantly greater thickness compared to the underlying layers. Without experiencing remelting of the last layer, Samples 2 and 3 similarly exhibit thicker uppermost layers relative to their underlying layers. Additionally, the remelting process induces molten material flow; the uppermost layer shows greater convexity relative to the underlying layers.

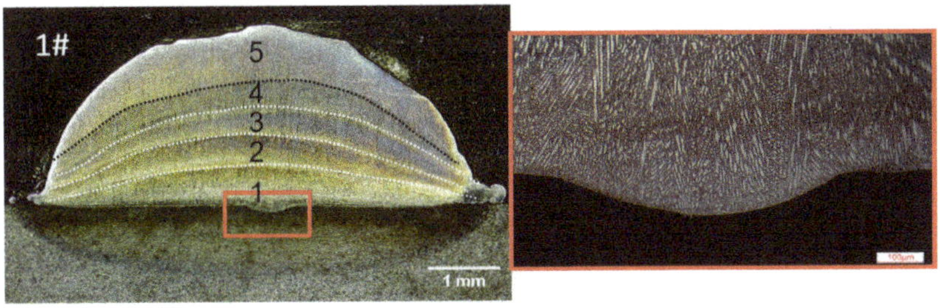


(a)

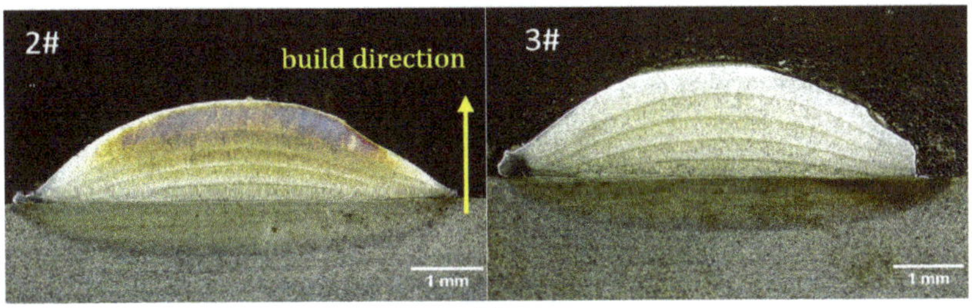


(b)

**Figure 4.** Cross-sectional morphology of the single-track multi-layer cladding. (**a**) Fusion lines and metallurgical bonding of Sample 1. (**b**) Cross sections of Samples 2 and 3.

Depending on the process parameters, the samples exhibited significant variations in both total build height and individual layer thickness. Figure 5 comparatively presents the layer-wise thickness and the standoff distance of the laser head during each layer deposition. In Figure 5a, due to the lower scanning speed, each individual layer belonging to Sample 1 displays a greater thickness compared to equivalent layers in the other samples. Especially the first and fifth layers of Sample 1 demonstrate significant thickness, exceeding their counterparts in Sample 3 by 127% and 89%, respectively. Samples 2 and 3 share identical process parameters (P and V), exhibiting slight layer thickness discrepancies due to different axis lifting amounts ($Z_d$).

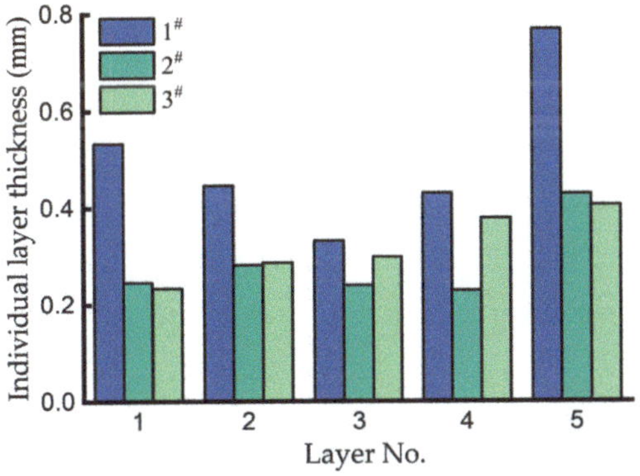


(a)

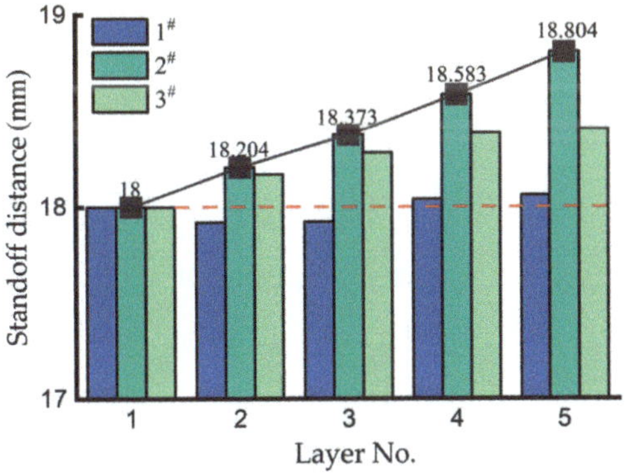


(b)

**Figure 5.** Individual layer thickness and the laser head's standoff distance for each layer deposition. (**a**) Comparative analysis of layer thickness. (**b**) Comparative analysis of the laser head's standoff distance (the red dashed line represents the initial standoff distance of 18 mm).

To maintain a consistent standoff distance, the laser head needs to be lifted a specified distance before the current layer's deposition. However, after the first layer, the thicknesses of subsequent layers are difficult to predict; thus, the first layer height has to be adopted as $Z_d$ and be kept consistent in subsequent layer deposition. $Z_d$ of Sample 1 was set to 0.45 mm by pre-estimation, while the $Z_d$ strategy was not used in Samples 2 and 3. As illustrated in Figure 5b, all layers in Sample 1 maintained standoff distances close to 18 mm. Without the $Z_d$ strategy corresponding to an increase in V, the standoff distances of Sample 2 became greater with the increase in the as-built layers, reaching 18.804 mm at the last layer. The laser beam and powder flow failed to concentrate on the focal plane, resulting in a continuous decrease in layer thickness, especially from the second to fourth layers. By adjusting $Z_d$ to 0.4 mm, Sample 3 demonstrated improvements in both layer thickness and standoff distances; however, it still exhibited undesirable deviations from the expected values. This indicates that accurate axial lifting amount estimation is critical for improving forming accuracy.

#### 3.2.2. Layer-Wise Width

The initial deposition layer undergoes more rapid cooling because of the substrate's heat dissipation effect, while it demonstrates progressively slower cooling rates due to thermal accumulation from preceding layers. Gravity-induced flow spreads molten material toward both edges of the cladding layer, resulting in wider deposition for subsequent layers compared to the initial layer. As the layers increase, a thin wall with a narrower base and a wider top is formed, presenting compromising dimensional precision.

Figure 6 presents a comparison of the results for the widths of the first and fifth layers within each sample layer. The results demonstrate a consistent pattern of width expansion in the fifth layer compared to the initial layer, with particularly significant increases observed in Samples 2 (38.5% growth rate) and 3 (34.7% growth rate). The width difference between the top and bottom layers of Sample 1 is relatively small, which might be due to the estimated axial lifting amount (i.e., the $Z_d$ strategy) having improved the geometric accuracy.

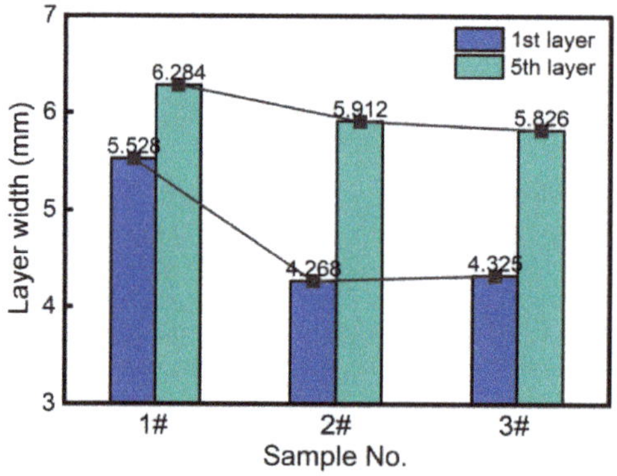


**Figure 6.** The width of the first and fifth layers of the samples.

### 3.3. *Microstructure and Mechanical Properties*

#### 3.3.1. Microstructure Analysis

It is generally recognized that the temperature gradient (G) and solidification rate (R) are pivotal parameters governing the grain morphology and dimensions in LDED [40]. The G/R ratio decides the grain morphology, and a high G/R ratio usually generates a planar grain; as R increases, the grains gradually transform to cellular, columnar and equiaxed grains. Moreover, G × R quantitatively defines the cooling rate which governs the grain sizes. Specifically, lower cooling rates promote coarse grain formation due to prolonged dendritic growth, whereas higher cooling rates induce grain refinement through rapid heterogeneous nucleation, contributing to enhanced mechanical properties [41].

Figure 7 presents the microstructure of different regions across the cladding layer. As illustrated in Figure 7b, a metallurgical belt formed by planar grains can be found at the cladding layer bottom. The planar grains exhibit epitaxial growth along the bonding interface, with needle-like crystals penetrating the substrate (marked by red arrow), forming a characteristic intergranular eutectic structure. This indicates good metallurgical bonding between the cladding layer and the substrate [9]. Near the bonding interface, the planar grains gradually transition into columnar and dendrite structures. This is mainly attributed to the temperature gradient (G) descending from the substrate to the cladding layer; non-equilibrium interfacial solidification usually induces grain morphology transformation [42]. During multi-layer cladding, the interlayer temperature gradient exhibits a continuous decline, while heat accumulation causes a corresponding deceleration in the solidification rate, ultimately leading to a reduction in the G/R ratio. Figure 7c,d present the microstructures within the middle and top regions of the cladding layer. Both dendritic and equiaxed grains are observable, with the top region of the cladding layer exhibiting a finer microstructure.

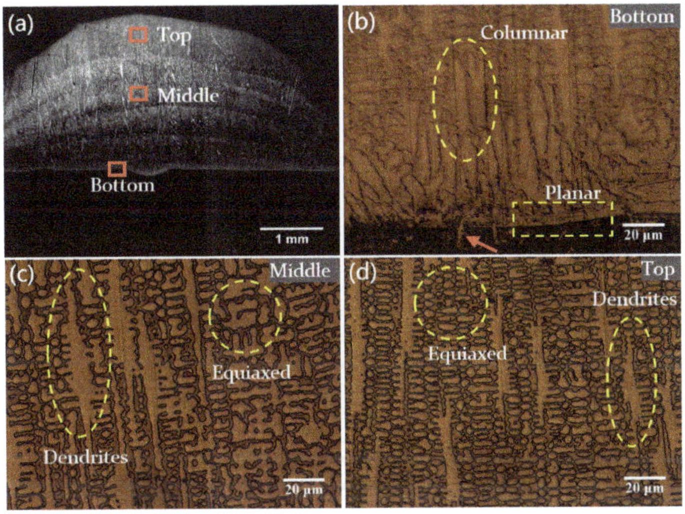


**Figure 7.** Microstructure of the cladding layer. (**a**) Different regions of Sample 1 in a cross section for microanalysis. (**b**) Bottom region of the layer. (**c**) Middle region of the layer. (**d**) Top region of the layer.

### 3.3.2. Microhardness Analysis

Microhardness tests were performed on the cross section of Sample 1. As shown in Figure 8, the microhardness varies significantly across different regions. Notably, the maximum value of approximately 900 $HV_{0.5}$ can be observed at the layer surface, while the substrate displays the minimum hardness value of merely 300 $HV_{0.5}$. The surface hardness of the cladding layer exhibits a remarkable enhancement, approximately three times higher than that of the substrate. The high microhardness is attributed to the finer grains formed under the low G/R ratio [43]. This substantial improvement conclusively demonstrates the effectiveness of the Fe-based cladding layer in significantly strengthening the mechanical properties of the substrate.

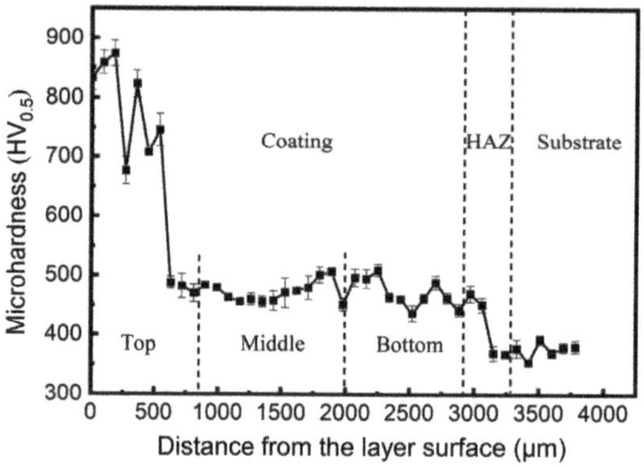


**Figure 8.** Cross-sectional mircohardness curve of Sample 1 (averages ± σ). The test direction is perpendicular to the bonding interface toward the substrate.

The middle and bottom regions of the cladding layer exhibit a relatively constant microhardness distribution, with slight fluctuations at the interlayer. This is probably attributable to the grain refinement of the fusion lines [39], exhibiting a distinct hardness enhancement. The coating layer, HAZ and the substrate can be clearly distinguished according to the microhardness curve. The high hardness of the layer surface greatly enhances its wear resistance, whereas the interior layer shows a uniform dendritic microstructure with a homogenous hardness distribution. The results confirm that the selected process parameters are well-optimized, contributing to superior cladding layer performance.

#### 3.3.3. Wear and Friction Analysis

Comparative results for wear morphology reveal differences between the substrate and the cladding layer in Figure 9. Due to the lower hardness, the substrate demonstrates significantly increased wear width and depth, exhibiting irregular abrasive tracks along the friction direction with pronounced surface deformation. The 3D topography exhibits a non-uniform wear depth distribution on the substrate surface, with the maximum wear depth reaching 9.1 μm, exceeding about 50% of the coating's wear depth.

The friction between abrasive materials and metal surfaces typically manifests as dry friction, with wear primarily occurring through material transfer and accompanied by the formation of wear debris. Furthermore, the hard abrasive particles mix into the metal surface under the normal force and consequently move along the tangential direction of the parallel force. This caused shear deformation and winkled morphology on the substrate surface. Plastic deformation displaces material within the plough laterally, resulting in accumulations forming characteristic ridges at groove boundaries. Under repeated friction, fatigue fracture and adhesion to the metal surface occurred. This indicates that substrate wear predominantly comprises pronounced adhesive wear and abrasive wear mechanisms. As for the coating layer, embedded wear debris and abrasive particles performed micro-cutting upon being pressed into the surface. Therefore, Figure 9b primarily presents ploughing grooves.

Comparative results for the friction coefficient between the substrate and cladding layer are presented in Figure 10. In the initial friction stage (0–400 s), the substrate friction coefficient rapidly increases from 0.2 to 0.87, while the coating friction coefficient gradually increases from near 0 to 0.5. During the intermediate phase (400–1500 s), the substrate maintains a stable high-friction state at approximately 0.92 with minimal fluctuations, while the coating displays a pronounced fluctuation amplitude as its coefficient slowly ascends to 0.6. In the final stage (>1500 s), both materials reach equilibrium states—the substrate stabilizes at ~0.9, the coating at ~0.56.

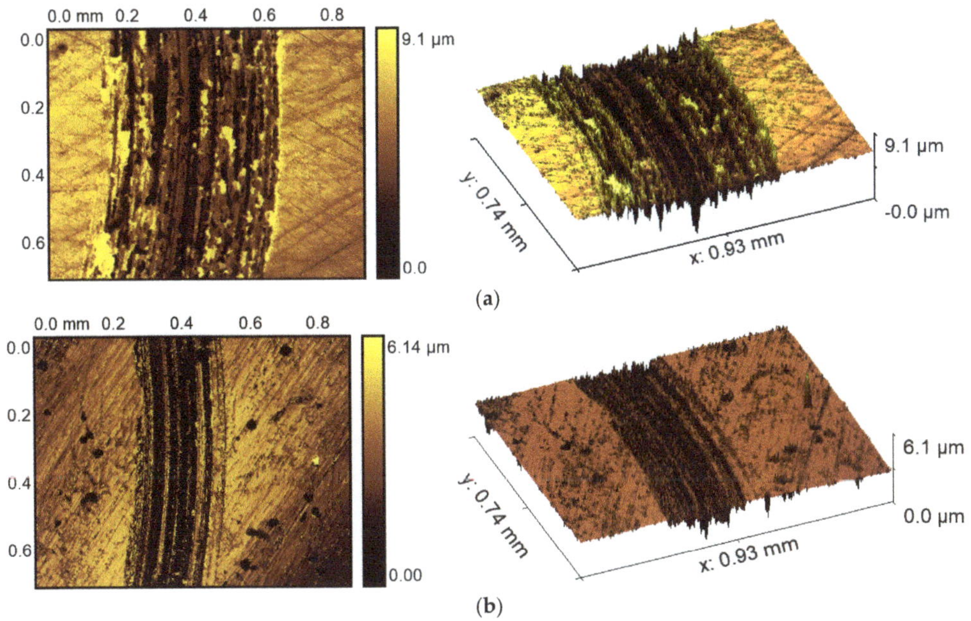


**Figure 9.** Wear morphology of the substrate and cladding layer. (**a**) Wear tracks and 3D morphology of the substrate. (**b**) Wear tracks and 3D morphology of the cladding layer.

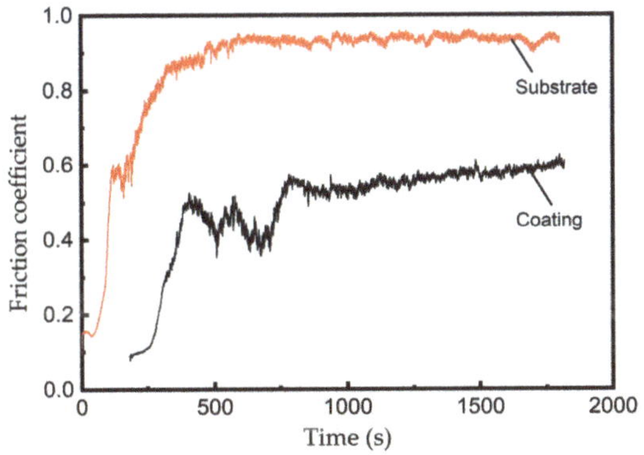


**Figure 10.** Friction coefficient–time curves of the substrate and cladding layer (the variance of the three repeated tests is 0.002).

The cladding layer shows a markedly lower friction coefficient than the substrate. Within a short duration, the substrate surface was sheared with grooves by the hard materials of the friction pair. The plastically deformed material adhered to the surface and was subsequently worn off, resulting in an initial decrease in the friction coefficient within a few minutes. As adhesion progressed, a stable friction regime was established, resulting in the cessation of friction coefficient fluctuations. Abrasive particles broke apart and embedded in the hard coating surfaces, resulting in a rapid increase in the friction coefficient. Then, as these abrasive particles penetrated deeper and were gradually worn smooth, the friction coefficient decreased. With repeated cycles, the furrow count grew and stabilized, explaining the early-stage friction coefficient fluctuations of the cladding layer.

## 3.4. *Cooling Rate Regulation and Microhardness Improvement*

### 3.4.1. Cooling Rate Analysis

Figure 11a presents the infrared temperature curves of the Fe-based alloy multi-layer cladding progress. The bead was 50 mm long, and a location (M) was chosen in the bead middle. The infrared images of each layer were recorded, and the radiation temperature of M was extracted. As the melt pool passed, M experienced a rapid temperature rise and consequent cooling down, with an interlayer cooling interval of 17 s. Due to heat accumulation, the first layer had the lowest temperature at the end of cooling. The temperature of the subsequent layers rose gradually; the last layer was 67 °C higher than the initial layer (see dashed circles in the Figure 11b). As can be seen in Figure 11b, the transient cooling rate of each cladding layer was computed through analysis of the temperature descent phase, using a sampling time of 120 ms. The cooling rate initially had a significantly high value, corresponding to the steep temperature curve, and then decreased progressively. It should be noted that point M is located at the weld bead centerline (along the scanning direction). As the melt pool passed though, point M's thermal history temperature was higher than that of the other positions far from the melt pool center, thereby inducing spatial variations in cooling rates. To systematically investigate the effects of scanning speed and interlayer dwell time on cooling rates, this study designated point M as the representative measurement location.

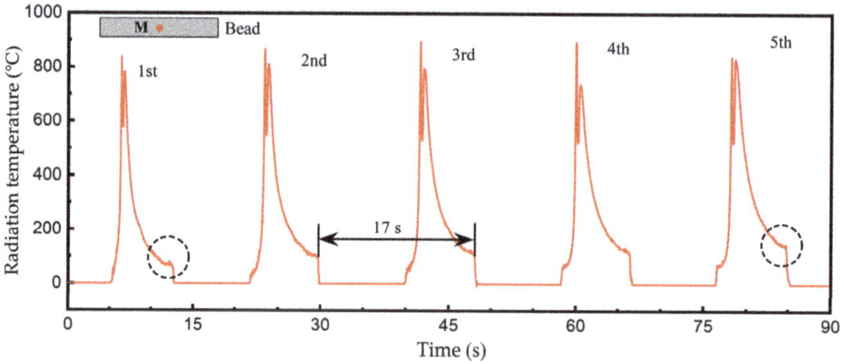


(**a**)

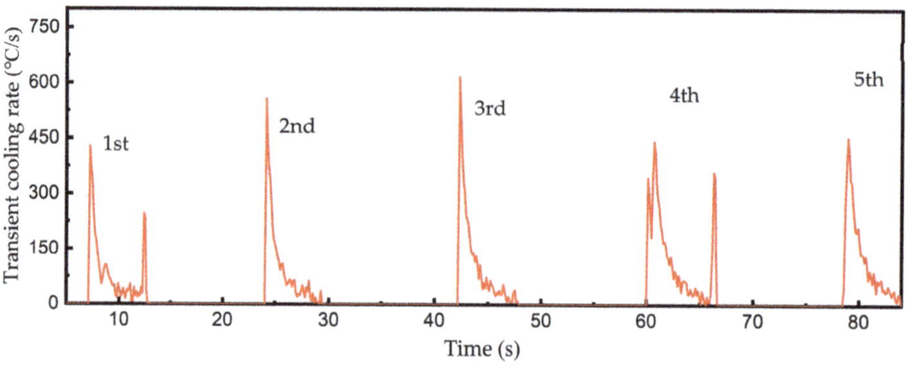


(**b**)

**Figure 11.** Thermal history and cooling rate of a fixed point (M) on the melt pool surface in multi-layer cladding processing (Sample 1). (**a**) Radiation temperature profiles of M on each cladding layer. (**b**) Calculated transient cooling rate of M according to the radiation temperature.

The transient cooling rate demonstrates a significant downward trend owing to the rapid temperature decrease, while showing high sensitivity to both measurement errors and infrared image disturbances. Furthermore, higher sampling frequencies tend to amplify the fluctuations observed in the cooling rate measurements. Therefore, the mean cooling rate of each cladding layer was computed and listed in Table 5.

**Table 5.** Mean cooling rate of each cladding layer.

| Layer No. | 1st | 2nd | 3rd | 4th | 5th |
|---|---|---|---|---|---|
| Mean cooling rate (°C/s) | 139 | 131 | 116 | 100 | 113 |

According to our developed thermal field simulation model [44], the cooling rates of various locations across the melt pool cross section were calculated for P = 2000 W and V = 7 mm/s. As shown in Figure 12, the initial high transient cooling rate and fluctuations present similar trends in the experimental results (Figure 11b). The melt pool surface demonstrates the highest initial cooling rate due to strong temperature gradients, while the interior regions present a substantial decrease in the cooling rate owing to the weakened convection. The cooling rates throughout the melt pool progressively converge to a stable value of approximately 130 °C/s, consistent with the experimental results. The analytical findings demonstrate that monitoring the melt pool surface cooling rate via online infrared imaging provides an effective indirect assessment of internal solidification dynamics.

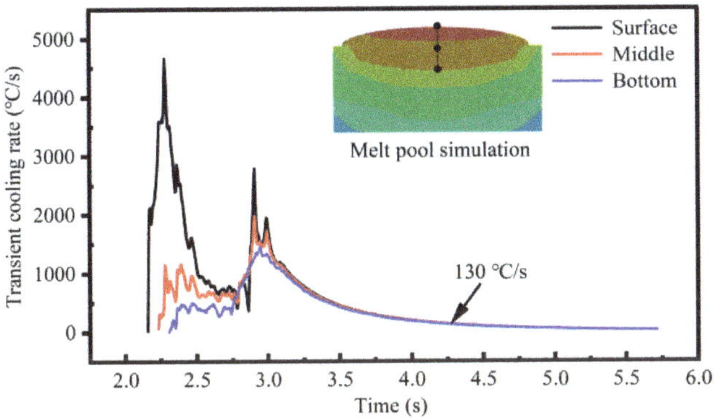


**Figure 12.** Cooling rate inside the melt pool computed through a thermal simulation model (P = 2000 W, V = 7 mm/s).

### 3.4.2. Cooling Rate Regulation

Modification of process parameters to regulate the cooling rate represents an effective methodology for controlling the microstructure and resultant properties of the cladding layer [40]. With the same interlayer dwell time ($T_I$) (5 s), the scanning speed (V) was increased to 12 mm/s for Sample 2 relative to Sample 1. Although the interlayer cooling interval was reduced from 17 s to 13 s, the mean cooling rate significantly increased, as shown in Table 5 and Figure 13a. The cooling rate reached 151 °C/s; the fourth layer increased by 45% compared to Sample 1. This indicates that the laser scanning speed predominantly affects the cooling rate. As the scanning speed increases, the laser line energy density (P/V) decreases and the melt pool interaction time drops, leading to faster solidification and a significant increase in the cooling rate. The reduced interlayer cooling interval provides a more insulating substrate for the subsequent cladding layer, which not only stabilizes the cooling rate but also substantially mitigates stress-induced cracks.

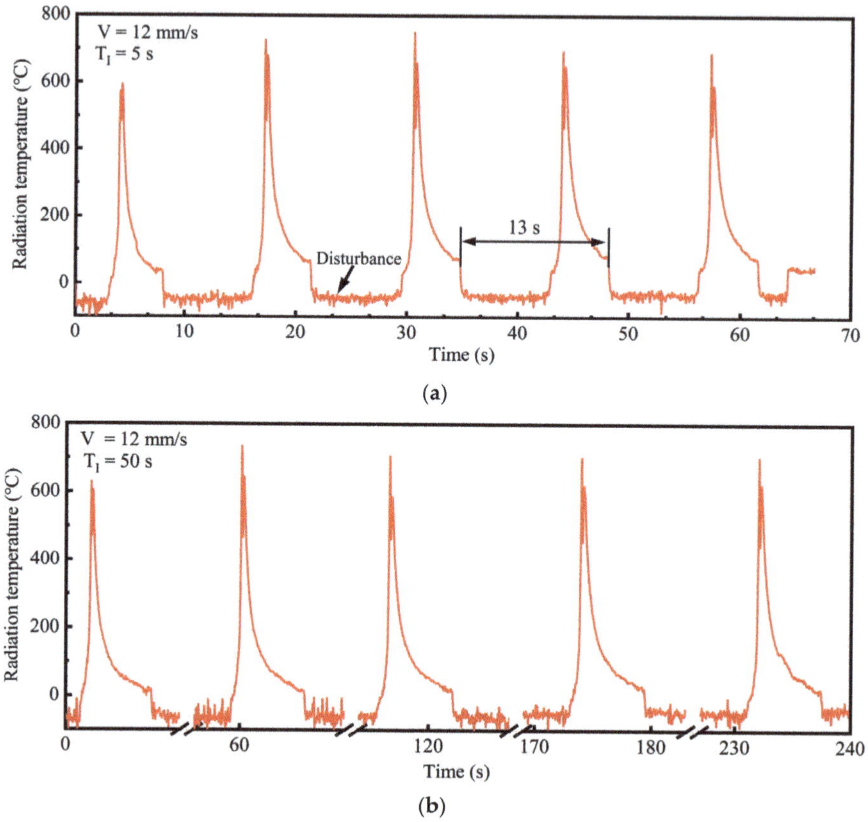


**Figure 13.** Thermal history and mean cooling rates of the melt pool surfaces. (**a**) Radiation temperature of Sample 2 layers, scanning speed (V) = 12 mm/s and interlayer dwell time ($T_I$) = 5 s. (**b**) Radiation temperature of Sample 3 layers, scanning speed (V) = 12 mm/s and interlayer dwell time ($T_I$) = 50 s.

For Sample 3, the interlayer dwell time was extended to 50 s while maintaining the same scanning speed as Sample 2. Figure 13b shows that the thermal history profiles demonstrate minimal variations across all the layers. The mean cooling rate exhibits a slight decrease because heat accumulation descends according to the thermal gradients. Notably, the last layer almost shows no change in the cooling rate (Table 6). The findings indicate that extending the interlayer dwell time mainly weakens the insulation effect of the cladding process while marginally enhancing the melt pool surface cooling rate. In addition, the cooling rate at ambient temperature mainly depends on the substrate bulk and thermal conduction; thermal radiation and convection between the cladding layer and the ambient environment exhibit a limited impact on cooling rates. To accelerate the cooling process, active cooling strategies have been widely used in recent years, including substrate in situ cooling [28] and immersion cooling techniques [45].

**Table 6.** Mean cooling rates of Samples 2 and 3.

| Layer No. | 1st | 2nd | 3rd | 4th | 5th |
|---|---|---|---|---|---|
| Mean cooling rate of Sample 2 (°C/s) | 151 | 155 | 148 | 145 | 129 |
| Mean cooling rate of Sample 3 (°C/s) | 130 | 129 | 117 | 129 | 128 |

A comparative analysis of the microstructure characteristics among the samples is illustrated in Figure 14. At a lower scanning speed (7 mm/s) and shorter interlayer dwell time (5 s), Sample 1 exhibits prolonged laser-material interaction. The laser line energy density (P/V) increased the melt pool temperature, extending the grain growth duration and resulting in significantly coarser grains (Figure 14c). In Figure 14b, the middle region of Sample 1 experiences an insulation effect, reducing the temperature gradient. This thermal condition promotes grain refinement while avoiding excessive cooling. Conversely, the top region demonstrates higher cooling rates and undercooling degrees than the middle region, further enhancing grain refinement (Figure 14a). When the scanning speed increases from 7 mm/s to 12 mm/s, P/V reduces significantly and results in a temperature decrease and solidification acceleration of the melt pool. The resulting high cooling rate effectively suppresses grain growth. Therefore, Samples 2 and 3 exhibit significant grain refinement across all regions relative to Sample 1, with the most pronounced effect observed at the bottom of the cladding layer (see Figure 14c,f,i). Extending the interlayer time from 5 s to 50 s, the second to fifth layers of Sample 3 exhibit longer cooling times. This inhibits the heat accumulation and promotes the cooling rate. The bottom region of Sample 2 shows the same microstructure characteristics as Sample 3, while the middle region has finer grains corresponding to the higher cooling rate (Figure 14e,h). However, the top region of Sample 3 displays a finer microstructure than Sample 2, indicating that prolonging the interlayer dwell time can effectively enhance equiaxed dendritic crystal formation and refinement.

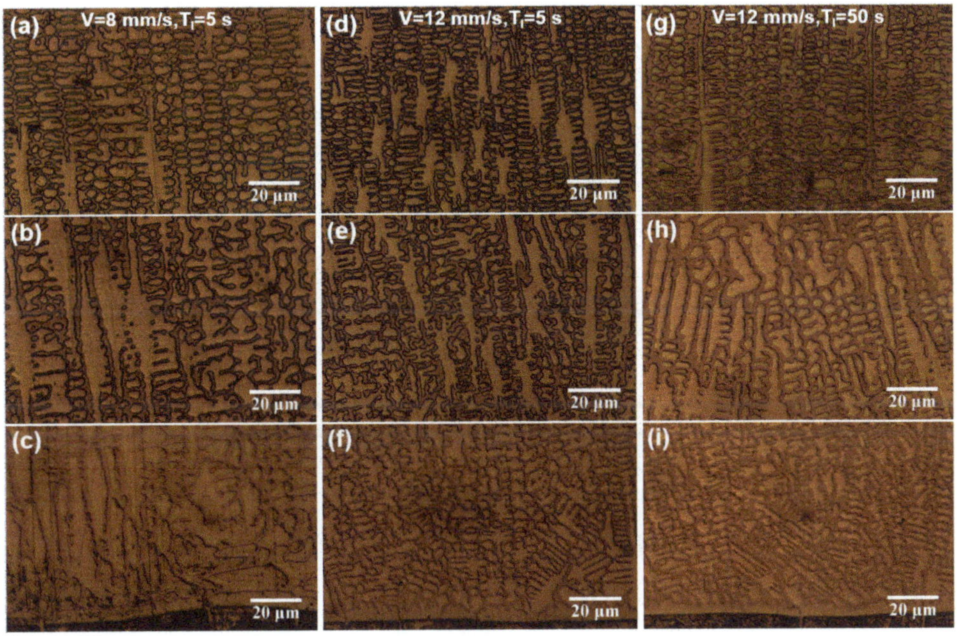


**Figure 14.** Comparative analysis of the effect of microstructure on cooling rate regulation. (**a**,**d**,**g**) Top regions of Samples 1~3. (**b**,**e**,**h**) Middle regions of Samples 1~3. (**c**,**f**,**i**) Bottom regions of Samples 1~3.

As shown in Figure 15, the microhardness values of Samples 2 and 3 exceed 500 $HV_{0.5}$ throughout the cladding layers, indicating significant hardness enhancement compared to Sample 1 (Figure 8). Notably, the top region of Sample 3 shows a higher hardness than Sample 2. The results coincide well with the microstructure characteristics of the samples in Figure 14.

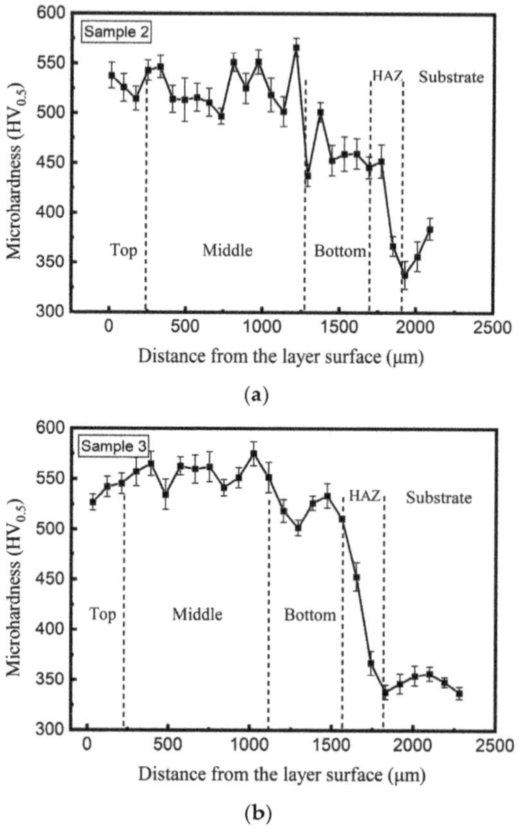


**Figure 15.** Comparative analysis of the effect of microhardness on interlayer dwell time (average ± σ). (**a**) Microhardness of Sample 2. (**b**) Microhardness of Sample 3.

Based on the performed analysis, the cooling rate can be effectively increased by either elevating the scanning speed or extending the interlayer dwell time, which consequently leads to significant improvements in both hardness and wear resistance properties. It should be emphasized that enhanced cooling rates may increase the crack susceptibility of the cladding layer, potentially inducing defects. However, Fe-based alloy materials have low crack sensitivity; there were no defects found in the samples, indicating the reliability of cooling rate enhancement. Furthermore, there were still difficulties in terms of closed-loop control of the transient cooling rate, and the underlying relationship between the cooling rate and the microstructure and properties of the cladding layer still requires in-depth research.

## 4. Conclusions

In this work, single-track muti-layer deposition experiments were conducted to study the geometric and microstructure characteristics of Fe-based alloy cladding layers. The radiation temperature of the melt pool was recorded online via an infrared camera, and the transient cooling rate was calculated according to the thermal history. Comparative analyses were performed to confirm the effectiveness of cooling rate enhancement in improving layer hardness. The main findings are as follows.

(1) Interlayer fusion lines appear distinctly due to coarsened grains formed by remelting of the pre-built layer. The geometric accuracy can be significantly improved by virtue of a predicted axis lifting amount ($Z_d$), which ensures a constant standoff distance during each layer's deposition.
(2) The Fe-based cladding layer exhibits different grain morphologies and sizes throughout the cross section. Planar grains form at the interface between the substrate and the cladding layer, and coarse columnar grains can be observed perpendicular to the bonding interface. Both dendritic and equiaxial grains coexist in the middle and top regions of the cladding layer, with finer grains on the top.
(3) Increasing the laser scanning speed or extending the interlayer dwell time can effectively enhance the cooling rate, resulting in appreciable improvements in coating hardness and wear resistance. The former approach demonstrates greater efficiency and effectiveness in enhancing the cooling rate. For large-part fabrication or large-area cladding, the deposition path should be rationally planned to increase the interlayer dwell time without compromising production efficiency.

**Author Contributions:** Conceptualization, B.H.; Formal analysis, J.X.; Investigation, B.H.; Resources, J.W.; Writing—original draft, B.H.; Supervision, J.W.; Software, Q.W. and L.Z.; Funding acquisition, B.H. All authors have read and agreed to the published version of the manuscript.

**Funding:** This research was funded by the Science and Technology Program of Henan Province, grant number 252102220103.

**Institutional Review Board Statement:** Not applicable.

**Informed Consent Statement:** Not applicable.

**Data Availability Statement:** The original contributions presented in this study are included in the article. Further inquiries can be directed to the corresponding author.

**Conflicts of Interest:** The authors declare no conflicts of interest.

## ferences

John, M.; Kuruveri, U.B.; Menezes, P.L. Laser Cladding-Based Surface Modification of Carbon Steel and High-Alloy Steel for Extreme Condition Applications. *Coatings* **2022**, *12*, 1444. [CrossRef]
Liu, G.; Su, Y.; Pi, X.; Xin, S.; Li, K.; Liu, D.; Lin, Y.C. Achieving high strength 316L stainless steel by laser directed energy deposition-ultrasonic rolling hybrid process. *Mater. Sci. Eng. A* **2024**, *903*, 146665. [CrossRef]
Ahn, D. Directed Energy Deposition (DED) Process: State of the Art. *Int. J. Precis. Eng. Manuf.-Green Technol.* **2021**, *8*, 703–742. [CrossRef]
Yao, M.; Kong, F. A review on laser cladding with wire feeding: Process fundamentals, theoretical analyses, online monitoring, and quality controls. *Int. J. Adv. Manuf. Technol.* **2025**, *137*, 4209–4242. [CrossRef]
Li, S.; Kumar, P.; Chandra, S.; Ramamurty, U. Directed energy deposition of metals: Processing, microstructures, and mechanical properties. *Int. Mater. Rev.* **2023**, *68*, 605–647. [CrossRef]
Li, Z.; Sui, S.; Ma, X.; Tan, H.; Zhong, C.; Bi, G.; Clare, A.T.; Gasser, A.; Chen, J. High deposition rate powder- and wire-based laser directed energy deposition of metallic materials: A review. *Int. J. Mach. Tools Manuf.* **2022**, *181*, 103942. [CrossRef]
Svetlizky, D.; Zheng, B.; Vyatskikh, A.; Das, M.; Bose, S.; Bandyopadhyay, A.; Schoenung, J.M.; Lavernia, E.J.; Eliaz, N. Laser-based directed energy deposition (DED-LB) of advanced materials. *Mater. Sci. Eng. A* **2022**, *840*, 142967. [CrossRef]
Liu, Y.; Ding, Y.; Yang, L.; Sun, R.; Zhang, T.; Yang, X. Research and progress of laser cladding on engineering alloys: A review. *J. Manuf. Process.* **2021**, *66*, 341–363. [CrossRef]
Liu, L.; Lin, Y.; Peng, L.; Kang, X.; Wang, X. Progress in Microstructure Design and Control of High-Hardness Fe-Based Alloy Coatings via Laser Cladding. *Coatings* **2024**, *14*, 1351. [CrossRef]
Wu, F.; Chen, T.; Wang, H.; Liu, D. Effect of Mo on Microstructures and Wear Properties of In Situ Synthesized Ti(C,N)/Ni-Based Composite Coatings by Laser Cladding. *Materials* **2017**, *10*, 1047. [CrossRef]
Shang, F.; Xu, Z.; Chen, J.; Chen, L.; Lu, S.; Wang, S.; Jing, X.; Dubovyi, O. Effect of Cr on wear resistance and mechanism of preset-powder laser cladding T15 alloy steel coating. *J. Mater. Res. Technol.* **2024**, *29*, 4991–4999. [CrossRef]

12. Kong, L.; Shi, C.; Hao, X.; Wang, S.; Huo, Y.; Zhu, Z.; Sun, J. Effect of Ti and Mo content changes on microstructure and properti of laser cladding FeCoCrNiMn high entropy alloy coatings. *Intermetallics* **2024**, *175*, 108509. [CrossRef]
13. Shang, F.; Chen, S.; Zhang, C.; Liang, J.; Liu, C.; Wang, M. The effect of Si and B on formability and wear resistance preset-powder laser cladding W10V5Co4 alloy steel coating. *Opt. Laser Technol.* **2021**, *134*, 106590. [CrossRef]
14. Wang, H.; Cheng, Y.; Geng, R.; Wang, B.; Chen, Y.; Liang, X. Comparative study on microstructure and properties of Fe-bas amorphous coatings prepared by conventional and high-speed laser cladding. *J. Alloys Compd.* **2023**, *952*, 169842. [CrossRef]
15. Ribeiro, K.S.B.; Mariani, F.E.; Coelho, R.T. A Study of Different Deposition Strategies in Direct Energy Deposition (DED) Process *Procedia Manuf.* **2020**, *48*, 663–670. [CrossRef]
16. Shin, S.; Jeon, I.; Sohn, H. Enhanced geometric accuracy in directed energy deposition via closed-loop melt pool height contr using real-time thermal imaging. *Addit. Manuf.* **2025**, *109*, 104846. [CrossRef]
17. Wang, K.; Liu, W.; Hong, Y.; Sohan, H.; Tong, Y.; Hu, Y.; Zhang, M.; Zhang, J.; Xiang, D.; Fu, H.; et al. An Overview Technological Parameter Optimization in the Case of Laser Cladding. *Coatings* **2023**, *13*, 496. [CrossRef]
18. Pinkerton, A.J.; Li, L. The significance of deposition point standoff variations in multiple-layer coaxial laser cladding (coaxi cladding standoff effects). *Int. J. Mach. Tools Manuf.* **2004**, *44*, 573–584. [CrossRef]
19. Wang, X.; Deng, D.; Hu, Y.; Ning, F.; Wang, H.; Cong, W.; Zhang, H. Overhang structure and accuracy in laser engineered n shaping of Fe-Cr steel. *Opt. Laser Technol.* **2018**, *106*, 357–365. [CrossRef]
20. Mazzarisi, M.; Errico, V.; Angelastro, A.; Campanelli, S.L. Influence of standoff distance and laser defocusing distance on dire laser metal deposition of a nickel-based superalloy. *Int. J. Adv. Manuf. Technol.* **2022**, *120*, 2407–2428. [CrossRef]
21. Zhu, G.; Li, D.; Zhang, A.; Pi, G.; Tang, Y. The influence of standoff variations on the forming accuracy in laser direct met deposition. *Rapid Prototyp. J.* **2011**, *17*, 98–106. [CrossRef]
22. Yu, T.; Sun, J.; Qu, W.; Zhao, Y.; Yang, L. Influences of z-axis increment and analyses of defects of AISI 316L stainless steel hollo thin-walled cylinder. *Int. J. Adv. Manuf. Technol.* **2018**, *97*, 2203–2220. [CrossRef]
23. Li, P.; Ji, S.; Zeng, X.; Hu, Q.; Xiong, W. Direct laser fabrication of thin-walled metal parts under open-loop control. *Int. J. Ma Tools Manuf.* **2007**, *47*, 996–1002.
24. Hu, B.; Han, J.; Wang, J.; Li, X. Prediction of the bead profile for wide-band laser cladding via hybrid model of mathematics ar thermal simulation. *Opt. Laser Technol.* **2024**, *170*, 110206. [CrossRef]
25. Shan, B.; Chen, J.; Chen, S.; Ma, M.; Ni, L.; Shang, F.; Zhou, L. Laser cladding of Fe-based corrosion and wear-resistant allc Genetic design, microstructure, and properties. *Surf. Coat. Technol.* **2022**, *433*, 128117. [CrossRef]
26. Liu, W.; Li, L.; Mi, G.; Wang, J.; Pan, Y. Effect of Fe Content on Microstructure and Properties of Laser Cladding Inconel 625 All *Materials* **2022**, *15*, 8200. [CrossRef]
27. Khodabakhshi, F.; Farshidianfar, M.H.; Gerlich, A.P.; Khajepour, A.; Mohammadi, M.; Withers, P.J. Engineering of textur gradient microstructures using directed energy deposition: The impact of adaptive cooling rate. *Mater. Design* **2024**, *245*, 1132 [CrossRef]
28. Reindl, T.; Hempel, N.; Mayr, P. Effect of substrate preheating and in situ cooling on dimensional accuracy, microstructure, ar hardness of wire-arc directed energy deposition components. *Weld World* **2025**. [CrossRef]
29. Zhang, H.; Mei, K.; Guo, W.; Li, Z.; Lai, Y.; Zhao, W.; Xiao, G.; Zhang, Y.; Cha, X. Comparative study on microstructures ar properties of air-cooled and water-cooled Fe-based plasma arc cladding layers. *J. Mater. Res. Technol.* **2023**, *23*, 1599–16 [CrossRef]
30. Mithal, A.; Maharjan, N.; Idapalapati, S. Enhanced cooling rates in laser directed energy deposition with interlayer peenin *Rapid Prototyp. J.* **2023**, *29*, 1289–1298. [CrossRef]
31. Foster, B.K.; Beese, A.M.; Keist, J.S.; McHale, E.T.; Palmer, T.A. Impact of Interlayer Dwell Time on Microstructure and Mechanic Properties of Nickel and Titanium Alloys. *Metall. Mater. Trans. A Phys. Metall. Mater. Sci.* **2017**, *48*, 4411–4422. [CrossRef]
32. Guévenoux, C.; Hallais, S.; Charles, A.; Charkaluk, E.; Constantinescu, A. Influence of interlayer dwell time on the microstructu of Inconel 718 Laser Cladded components. *Opt. Laser Technol.* **2020**, *128*, 106218. [CrossRef]
33. Chen, Z.; Zhang, Z.; Yang, Y.; Xiao, G.; Yi, M.; Zhou, T.; Xu, C. Influence of interlayer dwell time on microstructure ar mechanical properties additively manufactured 316L stainless steel by laser directed energy deposition. *J. Mater. Res. Techn* **2025**, *34*, 1304–1312. [CrossRef]
34. Farshidianfar, M.H.; Khajepour, A.; Gerlich, A.P. Effect of real-time cooling rate on microstructure in Laser Additive Manufact ing. *J. Mater. Process. Technol.* **2016**, *231*, 468–478. [CrossRef]
35. Nair, A.M.; Mahata, A.; Choudhury, A.R.; Nath, A.K. Effect of adaptive control of cooling rate on microstructure and mechanic properties of laser additively manufactured Inconel 718 thin walls. *Mater. Charact.* **2024**, *213*, 114032. [CrossRef]
36. Hu, B.; Han, J.; Wang, J. Parameter Study and Economic Efficiency Optimization for Laser Cladding with Wide-Band Fiber Las *Int. J. Opt.* **2022**, *2022*, 6373772. [CrossRef]
37. Farshidianfar, M.H.; Khajepour, A.; Gerlich, A. Real-time control of microstructure in laser additive manufacturing. *Int. J. Ac Manuf. Technol.* **2016**, *82*, 1173–1186. [CrossRef]

Derekar, K.; Lawrence, J.; Melton, G.; Addison, A.; Zhang, X.; Xu, L. Influence of Interpass Temperature on Wire Arc Additive Manufacturing (WAAM) of Aluminium Alloy Components. *MATEC Web Conf.* **2019**, *269*, 5001. [CrossRef]
Liu, C.M.; Tian, X.J.; Tang, H.B.; Wang, H.M. Microstructural characterization of laser melting deposited Ti–5Al-5Mo–5V–1Cr–1Fe near β titanium alloy. *J. Alloys Compd.* **2013**, *572*, 17–24. [CrossRef]
Farshidianfar, M.H.; Khodabakhshi, F.; Khajepour, A.; Gerlich, A.P. Closed-loop control of microstructure and mechanical properties in additive manufacturing by directed energy deposition. *Mater. Sci. Eng. A* **2021**, *803*, 140483. [CrossRef]
Liu, Z.; Zhao, D.; Wang, P.; Yan, M.; Yang, C.; Chen, Z.; Lu, J.; Lu, Z. Additive manufacturing of metals: Microstructure evolution and multistage control. *J. Mater. Sci. Technol.* **2022**, *100*, 224–236. [CrossRef]
Xie, Y.; Chen, W.; Liang, L.; Huang, B.; Zhuang, J. Influence of laser power on the microstructure and properties of Fe314 alloy cladding layer on EA4T steel. *Weld World* **2022**, *66*, 1551–1563. [CrossRef]
Gao, W.; Wang, S.; Hu, K.; Jiang, X.; Yu, H.; Sun, D. Effect of laser cladding speed on microstructure and properties of titanium alloy coating on low carbon steel. *Surf. Coat. Technol.* **2022**, *451*, 129029. [CrossRef]
Hu, B.; Han, J.; Li, X.; Wang, J. Modeling of melt pool and thermal field simulation for wide-band laser cladding. *Opt. Eng.* **2023**, *62*, 116103. [CrossRef]
Costa, J.N.; de Assis Faria, G.; Porcaro, R.R.; Pereira, I.C. Evaluation of near immersion active cooling on the microstructure and mechanical properties of AISI 316L stainless steel obtained with additive manufacturing by DED-Arc. *Int. J. Adv. Manuf. Technol.* **2024**, *134*, 1419–1432. [CrossRef]

materials

MDPI

Article

# Investigation of Laser Macro- and Micro-Polishing on Fine-Grained Niobium Material for Superconducting Radio Frequency Cavities

**Florian Brockner [1], Laura Kreinest [2], Edgar Willenborg [2] and Dirk Lützenkirchen-Hecht [1,*]**

[1] Fakultät 4-Physik, Bergische Universität Wuppertal, Gaußstr. 20, D-42119 Wuppertal, Germany; brockner@uni-wuppertal.de

[2] Fraunhofer Institut für Laser Technology ILT, D-52074 Aachen, Germany; edgar.willenborg@ilt.fraunhofer.de (E.W.)

* Correspondence: dirklh@uni-wuppertal.de

**Abstract**

Fine-grained Nb metal sheets were successively laser macro- and micro-polished for a potential use of the so-prepared materials in superconducting radiofrequency cavities in particle accelerators. The laser-treated Nb surfaces were investigated by a combination of white light interferometry, optical profilometry, electron microscopy with X-ray spectroscopy, and X-ray diffraction to study the influence of the conditions during the laser treatments on the resulting surface topography, the crystallographic structure, and the chemical composition of the material samples. For optimum polishing conditions, smooth, wavy surfaces with a minimum surface roughness could be achieved. However, local defects such as carbon contamination, as well as holes and cracks in the surface, were found. For the different prepared surfaces, the maximum acceleration field gradients, i.e., the onset fields for field emission ($E_{On}$), were determined, indicating that for smooth surface regions without defects, $E_{On}$ may reach values of up to almost 1 GV/m, while for the pristine, rough surface and local defects such as particles and cracks, $E_{On}$ is limited to values around 100 MV/m or less. The present study suggests that laser polishing should be considered as an alternative to conventional polishing strategies of niobium accelerator cavities.

**Keywords:** niobium; laser polishing; superconductivity

## 1. Introduction

Almost all modern particle accelerators, such as the Large Hadron Collider at CERN [1], the European Spallation Source [2], or the European X-ray free-electron laser (X-FEL) [3], rely on superconducting radio-frequency (SRF) cavity resonators made of niobium. In order to enable the fabrication of high-performance cavities with sufficient accelerating field gradients of 30 MV/m or more, sophisticated processing techniques have been developed even for an application on an industrial scale. On a laboratory scale, values of about 50 MV/m, i.e., close to the theoretical limit [4], are feasible [5]. State-of-the-art manufacturing processes include chemical polishing to remove contaminations at the surface of the materials, and different heat treatments in high vacuum are used to reduce crystallographic defects and stress in the lattice (e.g., [5,6]). Elaborate polishing procedures, including buffered chemical etching in hydrofluoric acid and nitric acid [7], centrifugal barrier polishing, and electrochemical polishing, again employing HF [8], are used to transform the initially technically rough starting surfaces to a mirror-like finish, removing also the crystallographic defects induced by forging and mechanical polishing. These smooth surfaces are an intrinsic prerequisite to suppress parasitic electron field emission processes that may limit the applicable acceleration fields [9]. Last but not least, residues from the polishing solutions have to be removed from the cavity surfaces by additional cleaning steps subsequently, e.g., by high-pressure water rinsing [7,8] or dry ice cleaning [10].

It is important to note that current preparation routes for Nb cavities consume large volumes of toxic and harmful hydrofluoric acid for chemical polishing and electropolishing, and the heat treatments of the cavities are performed in ovens with wall-plug powers of several kW for many hours or even days. Those treatments are needed to remove hydrogen incorporated in the niobium material, which tends to form non-superconducting Nb-hydrides, and due to the contact with aqueous electrolytes, the formation of surface oxides is facilitated. In order to allow a more economic construction and operation of future accelerators, such as the planned Future Circular Collider [11] or the International Linear Collider [12], a further increase in the field strength and a reduction in the electric power dissipation are necessary, and, in accordance with the Paris Agreement [13], new, innovative, energy-efficient, and environmentally friendly preparation methods for the accelerators are needed. Facile, economic, and energy-efficient cavity processing techniques may also facilitate a more widespread use of accelerator technology, e.g., in industrial or medical applications.

In this context, dry laser polishing processes appear to be of potential interest: intense, pulsed, or continuous laser beams are used to melt the surfaces of the workpieces, effectively removing contaminations such as the above-mentioned hydrides and oxides and smoothing the machined surfaces [14–16]. Compared to the conventional cavity treatments, the efficient energy transfer to the near-surface region of the workpieces allows a substantial reduction in the applied energy, and any toxic or potentially harmful liquids are entirely avoided. However, the interactions of the laser beams with the materials of the workpiece are very complex and need to be optimized for each material individually [16–20]. However, laser polishing provides excellent opportunities for a tailored treatment; for example, energy regulation is easily possible by the variation in the laser power, the number of used laser pulses, repeating frequency, laser focus size, and the scan speed on the workpiece, so that the heat introduced to the material can be precisely controlled, and thus, any unwanted bulk heating of the material can be avoided. Surface cleaning processes are also possible employing laser treatments [21], for which commercial solutions are available [22]. In relation to laser treatments of cavity materials for accelerators, it was recently shown that scratches remaining after the conventional polishing processes can be removed purposefully, allowing for a substantial increase in accelerating field gradients and suppressing parasitic electron field emission [23,24].

Up to now, laser polishing processes have been restricted to laboratory-sized setups, as proof-of-principle studies were the focus of interest, and accordingly, the polished surface areas are of the order of several mm$^2$ only [25,26]. It is thus of high interest whether the approach can be transferred to a near-technology scale, with a fast treatment of larger surface areas. In the present study, therefore, we have applied laser macro- and micro-polishing to an untreated, fine-grained Nb sheet directly taken from the fabrication of the Tesla cavities for the X-FEL and studied in detail the effects of the laser polishing treatments on the resulting surface structures on the Nb material and provide evidence that laser polishing might indeed be used for the large-scale processing of Nb cavities.

## 2. Materials and Methods

### *2.1. Materials*

Pieces of ca. 21 × 21 $mm^2$ were cut from a 3 mm thick fine-grain niobium sheet with 85.25 to 86.8 grains per $mm^2$, taken from the series production of the TESLA cavities for the European XFEL by high-pressure waterjet cutting at DESY, Hamburg. The sheet with a surface roughness $R_a$ = 0.97–1.43 μm was used for laser polishing without any mechanical or chemical pre-polishing in order to examine the feasibility of dry laser polishing treatments as a replacement for the conventional polishing strategies applied so far.

### *2.2. Laser Polishing*

The laser polishing processes were all carried out employing the equipment available at the Fraunhofer ILT (Aachen, Germany) [27]. All treatments were performed under an inert Ar-gas atmosphere with oxygen and humidity levels well below 500 ppm. A galvanometric laser scanner (hurrySCAN30, ScanLab GmbH, Puchheim, Germany) was used to move the beam in a meandering pattern over the niobium surface to be polished, and a computer-controlled, 5-axis portal machine positions the workpieces in the beam. While a CW fiber laser with an infrared wavelength of 1075 nm (redPOWER, SPI Laser Ltd., Southhampton, UK) was used for the macro-polishing, a Q-switched disk laser (TruMicro 7050, Trumpf, Ditzingen, Germany) with a wavelength of 1030 nm and a maximum average laser power of 520 W was employed for subsequent micro-polishing steps. Both laser optics employed a circular shape with a flat-top intensity distribution. The focus of the two laser beams can be manipulated by a zoom telescope. All important polishing parameters, such as the laser power, the feed rate, the beam diameter, the track offset as well as the number of polishing cycles, were systematically varied during the experiments. More details of the laser polishing setups are given, e.g., in refs. [28–31]. For each polishing condition, a test field of 5 × 5 $mm^2$ was treated and subsequently analyzed with white light interferometry, optical profilometry, electron microscopy with energy-dispersive X-ray analysis (SEM/EDX), and X-ray diffraction, respectively. Some selected samples were laser-polished on larger areas of up to 15 × 15 $mm^2$. Up to now, in total, more than 50 different test fields have been prepared and analyzed.

### *2.3. Materials Analysis*

The surface topography of the investigated niobium surfaces was investigated employing white light interferometry (WLI; Nexview NX2 (Zygo Corporation, Middlefield, CT, USA), with lateral and vertical resolution of about 3 μm and 1 nm, respectively. WLI data of 1 $mm^2$ areas in the center of the laser-polished regions are analyzed with a profile filter so that the roughness of the sample is obtained as a function of the spatial wavelength, as described in more detail in ref. [28]. Furthermore, optical profilometry (OP) was used to further study the surfaces of the samples with an FRT Metrology MicroProf system (now Camtek, Bergisch Gladbach, Germany) with a lateral resolution of ca. 2 μm and a vertical resolution of ca. 3 nm. Scanning electron microscopy was carried out in a JEOL JSM-6510 system (JEOL JSM 6510, JEOL Ltd., Tokyo, Japan) equipped with a Thermo Fisher Scientific (Waltham, Massachusetts, USA) EDX system (Noran 7, silicon drift diode) with an energy resolution of about 120 eV, which allows detection of light elements. X-ray diffraction measurements were performed on a Malvern Panalytical (Almelo, the Netherlands) multipurpose diffractometer with Cu K$\alpha$ radiation and a multistrip detector in the Bragg–Brentano parafocusing geometry.

### 2.4. Parasitic Field Emission Studies

Field emission measurements of the pristine and the different laser-polished samples were performed in a dedicated measurement system, generally operated under ultrahigh vacuum [32]. The system has a fixed anode and a cathode holder with the Nb sample, which can be moved in a computer-controlled raster pattern. In the present study, truncated tungsten cones with a diameter of 150 μm were used to measure the onset fields for electron field emission and the resulting currents as a function of the position on the sample, locally and on average.

## 3. Results and Discussion

A pre-screening of the obtained surface topographies after laser polishing treatments was used to select an appropriate focus spot of the laser and a closely connected line spacing of the laser on the surface during laser polishing, both for macro- and micro-polishing. Variation in the laser spot size ($D_L$) included the range from 250 μm to 500 μm, and line spacings (track offset $d_y$) from 50 μm to 100 μm. Furthermore, the scan speed (line feed $v_s$) was varied between 25 mm/s and 100 mm/s for the macro-polishing (MP). Furthermore, the number of polishing scans ($n$) was also varied, with the scan direction being rotated by 90 degrees after each scan. Best results were obtained for $D_L$(MP) = 250 μm, $d_y$(MP) = 50 μm, and $v_s$(MP) = 25 mm/s, so these parameters were used for all subsequent laser macro-polishing treatments (see Supplemental Information, Figure S1). In particular, the largest investigated laser power and the lowest scan speed lead to a larger melt pool with a higher temperature, so that the applied heat may lead to a more pronounced smoothing of the treated surfaces.

Similarly, the micro-polishing (μP) parameters were first screened, with resulting focus diameter of $D_L$(μP) = 280 μm, $d_y$(μP) = 30 μm, and $v_s$(μP) = 1000 mm/s with a pulse repetition frequency of 20 kHz of 120 ns laser pulses. Some typical white light interferometry results are compiled in Figure 1, where the topography of the untreated niobium sheet (Figure 1a) is compared to that of a macro-polished area with a laser power of 240 W and a vertical and a second horizontal polishing scan (Figure 1b). Obviously, the roughness of the macro-polished sample is substantially reduced, i.e., the grooves from the mechanical processing of the samples can be efficiently removed by macro-polishing. However, although the surface contour appears much more homogeneous compared to the untreated sample, i.e., the polycrystalline nature of the sample becomes visible and a substantial reduction in surface roughness is obvious, the tracks remaining from the laser macro-polishing are clearly visible, despite the fact that two scans, one in a vertical and a second in a horizontal direction, have been performed (see, e.g., [29]). Those structures are a remainder of the localized melting of the material in the laser focal spot and its re-solidification, as schematically shown in Figure 1d. However, a further decrease in roughness and an increase in homogeneity can be noted after two additional micro-polishing treatments (Figure 1c).

The effects of multiple laser polishing treatments and the laser power are compiled in Figure 2. Here, the average roughness on different length scales as determined from a detailed analysis of the white light interferometry images is depicted for both the macro-polishing (Figure 2a) and the effect of an additional micro-polishing with a pulsed laser of different laser power (Figure 2b). As can be seen, a laser power of at least 210 W is required to induce a noticeable reduction in the roughness on a μm length scale after a single MP treatment (dark red curve in Figure 2a), and a further increase in the laser power to 240 W allows us to substantially reduce roughness on all length scales (blue curve in Figure 2a). If a second MP treatment is applied with $P_L$(MP) = 240 W, the roughness may be further reduced, i.e., by rotating the meander pattern by 90 degrees, a second MP treatment may reduce the tracks from the first polishing cycle as well as the remaining roughness features from the pristine sample (pink curve in Figure 2a). A third and even more MP cycles do not substantially alter the surface topography, as can be deduced from the light red curve in Figure 2a, so that all laser MP treatments were performed with two MP cycles and a laser power of 240 W (see also Figure S1 in the Supplementary Material). In the context of cavity production and the objective to suppress any parasitic field emission from the laser-polished surfaces, in particular, the roughness on short length scales is a key issue, and this is effectively reduced by a laser power of 240 W. For slow scan speed and high laser power, however, roughness on larger length scales in the order of mm increases, indicating that in this particular case, too much heat is introduced (see also Figure S1 in the Supplementary Material).

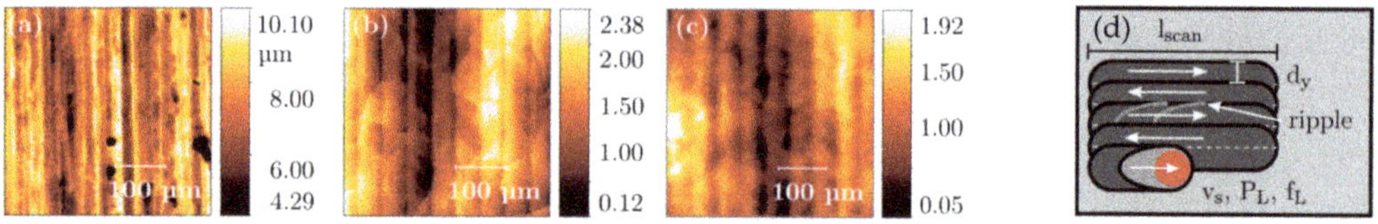


**Figure 1.** Topography of the niobium sheet investigated with white light interferometry prior to any polishing (**a**), after macro polishing ($P_L$(MP) = 240 W, one horizontal (h, x-direction) + one vertical (v, y-direction) scan) (**b**), and after an additional micro-polishing ($P_L$(µP) = 80 W, again two cycles in h and v) (**c**). (**d**) represents a scheme of how the rippled structure of the surface after laser polishing is generated in the x- and y-directions. The white arrows indicate the movements of the laser beam on the sample surface.

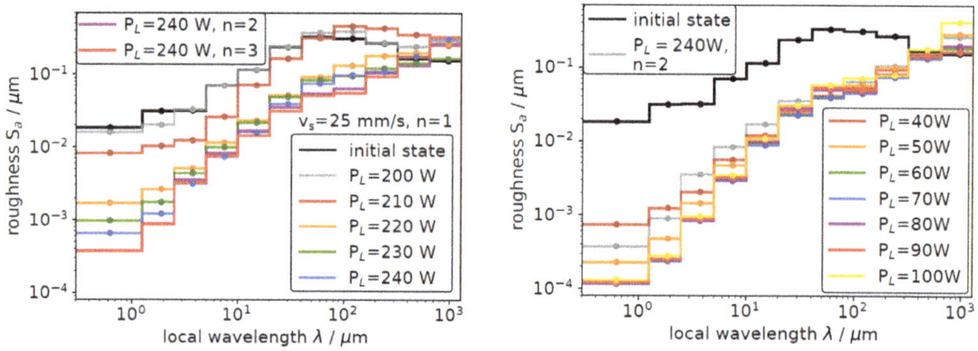


(**a**) (**b**)

**Figure 2.** Evaluation of the average surface roughness on representative areas of 1 mm$^2$ as a function of the lateral length scale as measured with light interferometry. (**a**) Roughness evaluation after macro-polishing for different laser powers, $P_L$, and different numbers of macro-polishing ($n$ = 1, 2, 3) at a speed of 25 mm/s. (**b**) Roughness evolution after an additional micro-polishing treatment for different average laser powers and a processing speed of 1000 mm/s in comparison to the initial surface topography of the untreated sample and the macro-polished sample ($P_L$(MP) = 240 W, $n$ = 2).

For a subsequently executed micro-polishing treatment, again, a threshold of at least 50–60 W laser power is necessary to further modify the topography of the macro-polished Nb samples (see Figure 2b). As can be seen, an average laser power of more than 90 W leads to an increase in surface roughness for larger local wavelengths, most probably by inducing stress in the treated metal sheet, induced by rapid heating and cooling cycles [29]. Optimum results are obtained for a laser power of $P_L$(µP) = 80 W and again, one vertical plus one horizontal meander scan.

In order to investigate the roughness features of the laser-treated surfaces in more detail, optical profilometry was applied, and in Figure 3, the results of the OP data evaluation are presented. More specifically, the power spectral density functions (PSDF), a measure of the roughness depending on the lateral frequency (wave number $k = 2\pi/\lambda$) [33,34]. In general, the PSDF represents the Fourier transform of the autocorrelation function $R(\xi)$ (Equation (1)),

$$R(\xi) = \lim_{L\to\infty} \frac{1}{L} \int_0^L z(x)z(x+\xi)dx. \quad (1)$$

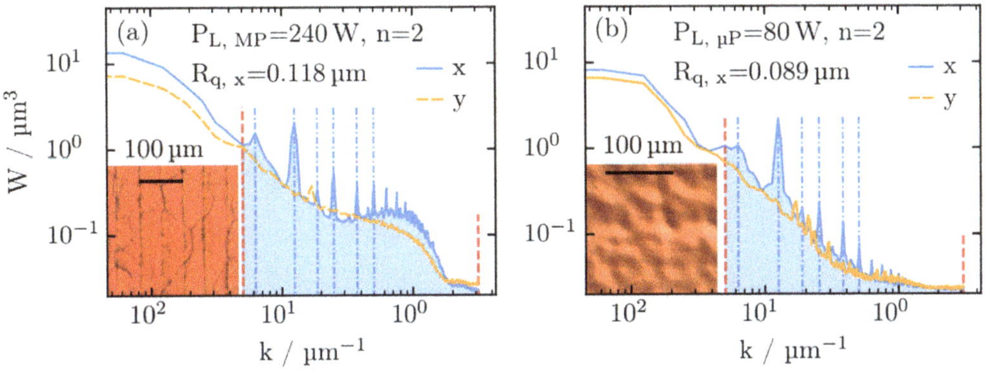


**Figure 3.** (**a**) Power spectral density functions (W) of an only macro-polished Nb sample ($P_L$(MP) = 240 W, n = 2), and (**b**) a second sample subsequently micro-polished ($P_L$(µP) = 80 W, $n$ = 2). The x- and y-directions correspond to a direction perpendicular and parallel to the last laser polishing treatment, respectively, and are treated separately. In the inserts, results of an inspection with light microscopy are shown for the two samples. The dashed blue lines correspond to peaks in the respective PSDFs at $\lambda$ = 13 µm, 17 µm, 25 µm, 33 µm, 50 µm, and 100 µm, and the red vertical lines present the k-range for which the average roughness values ($R_q$) are calculated ($\lambda_{min}$ = 2 µm, $\lambda_{max}$ = 120 µm).

$R(\xi)$ provides a measure of the probability of finding a point of identical height $z(x)$ at a distance $x + \xi$, and the PSDF is given by

$$W(k) = \frac{1}{2\pi} \int_{-\infty}^{\infty} R(\xi)\exp(-ik\xi)d\xi. \quad (2)$$

Thus, peaks in the PSDF correspond to periodic wave structures (wave number $k$) and are thus well-suited to identify traces of the tracks from the laser polishing as well as other laser-induced periodic structures on the surfaces [35]. Those periodic structures are not detectable by simply measuring the roughness as a function of their length scale, as depicted in Figure 2. Here we discriminate between the PSDF perpendicular to the direction of the last laser polishing scan (x-direction) and the PSDF parallel to the last laser polishing (y-direction).

In Figure 3a, the PSDF calculated after the macro-polishing ($P_L$(MP) = 240 W, $n$ = 2) is shown together with an optical inspection with a light microscope. While the PSDF is smooth along the polishing tracks and substantially decaying towards increasing wavenumber (decreasing length scale $\lambda$), distinct peaks are present in the x-direction. These peaks, located at $\lambda$ = 13 µm, 17 µm, 25 µm, 33 µm, 50 µm, and 100 µm, are the remainder of the polishing tracks and can also be seen in the light microscopy image in the inset of Figure 3a. The largest amplitude in the PSDF at 50 µm directly corresponds to the used line spacing $d_y$(MP) = 50 µm. For the direction along the polishing tracks (y-direction), any distinct peaks are absent, indicating that the melting and re-solidification of the niobium results in a reduced roughness of the laser-treated material already after the macro polishing.

After two additional micro-polishing treatments, the roughness further decreases in both the x- and y-directions; however, remnants of the frequency spectrum of the macro-polished sample still persist. Direct traces of the line spacing in the micro-polishing are expected for $\lambda = d_y(\mu P) = 30$ μm. However, still, the peak corresponding to the line spacing of the macro-polishing procedure is dominating the PSDF in the x-direction. Smaller peaks are also visible in the y-direction, which are a fingerprint of the overall smooth but wavy surface topography of the macro- and micro-polished sample. Here, no indications for line traces are detectable by light microscopy (insert of Figure 3b). The roughness calculated by integration of the PSDF from $\lambda_{min} = 2$ μm to $\lambda_{max} = 120$ μm along the x-direction, i.e., with an enhanced roughness, leads to a value of 0.118 μm for the macro-polished sample (Figure 3a) and only 0.089 μm for the macro- and micro-polished sample (Figure 3b). The most obvious decrease in roughness is observed in the length scale range below about 30 μm, which is most crucial for the suppression of parasitic electron field emission and breakdown of superconductivity in niobium cavity materials. In order to further reduce lines and ripples, a rotation of the sample during laser treatment and more sophisticated laser tracks on the polished workpiece could be considered (see Figure 1). Furthermore, the intensity distribution of the laser beam could be adapted.

An example of a full-field white-light interferometry image of a $5 \times 5$ mm$^2$ sample is shown in Figure 4a. Despite the fact that the entire surface was prepared with optimized laser polishing conditions, a large inhomogeneity can be stated. First of all, on a large scale, there are still some periodic features remaining from the macro-polishing, in the present case, in the up/down direction. Second, there are some remaining defects from the original niobium sheet, i.e., a vertical scratch from the rolling processes (black arrow in Figure 4a) and some pits marked by colored crosses. The depth of the pits was determined to be about 3.6 μm (green cross), 4.5 μm (orange cross), and 6.4 μm (blue cross), respectively. Obviously, laser polishing alone cannot remove all the different defects that are present on the pristine, original niobium sheet, and some mechanical or chemical pre-polishing appears to be required. For two line profiles, the average height variation was also determined, resulting in values of 8.7 μm (horizontal blue profile) and 7.7 μm (vertical orange profile), indicating that the sheet was slightly deformed or bent due to the laser polishing, or more material was evaporated from the center of the sample in the course of the laser processing.

In addition to the ripples, lines, and point defects, visible light microscopy proves the grain structure of the material, as well as the presence of cracks in the laser-treated material. The latter may appear because extremely high temperatures are needed for the melting of niobium, with a bulk melting temperature of 2477 °C (ref. [36], p. 51). Furthermore, for high-quality polishing results, the temperature should be as high as possible, i.e., close to the boiling point of Nb (4927 °C) [37]. Limited by heat dissipation to the environment of the sample, thereby the time for the existence of the melt is maximized, and polishing effects are optimized. However, heating and cooling occur on time scales of microseconds or less, so extreme heating and cooling rates result, and the thermal loads may induce mechanical stress in the material, finally causing cracks at the surfaces. Although these cracks are critical in view of the use of laser-polished samples in acceleration structures, crack formation may be avoided, e.g., by some moderate preheating of the laser-treated samples to slow down in particular the cooling processes, a further optimization of the laser processing, or an additional heat treatment after the laser polishing.

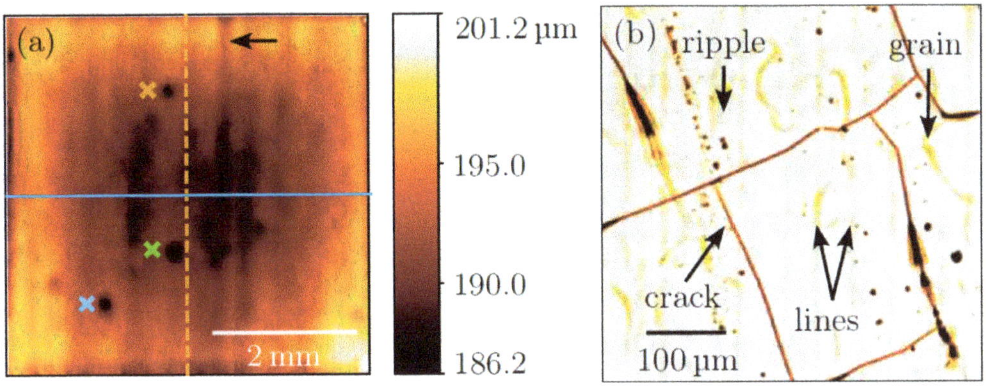


**Figure 4.** (**a**) White light interferometry of the entire surface of a laser macro- and micro-polished Nb sheet. Besides a wavy surface appearance, distinct defects are detectable, i.e., a remaining scratch (black arrow) and circular pits (colored crosses). Two line profiles (blue horizontal line and orange vertical line) are indicated, which were used to measure any macroscopic deformation induced by the laser polishing procedure. (**b**) Light microscopy of defects. Again, ripples as well as defects are detected and indicated accordingly. Traces of scan lines, as well as cracks in the surface and the grain structure of the material, are visible.

Further point defects detected after laser polishing appear to be related to chemical contamination. Although the specification for the pristine material only allows 2 µg carbon/g niobium, electron-excited X-ray analysis proves substantial amounts of C on the surface of the pristine and the laser-treated Nb (see Figure S2 in the Supplementary Material). While the average concentration determined for oxygen slightly decreases after the laser polishing, indicating the effectiveness of the protecting noble gas atmosphere, the concentration of carbon substantially increases, reaching a value of more than 7 atomic %. This can be explained by the presence of carbon inclusions in the Nb material that have a smaller density than metallic niobium and thus float on the liquid metal during the laser melting and reside at the surface after re-solidification. Examples of such carbon contaminations in the laser-polished sample and in the pristine, untreated material are shown in Figure 5a,b. According to the SEM analysis, the size of both defects is on the order of 20 µm, and the particle detected after laser polishing is clearly mainly composed of carbon. A detailed analysis typically provides carbon concentrations of 30–40 at. %, and 15–30 at. % oxygen, while substantially lower concentrations are found on average. On a particle-free Nb surface, typically only about 10–15 at. % C and less than 10 at. % O were detected.

As can be seen in the SEM micrograph of the untreated, pristine Nb sheet, several smaller particles are present on the surface of the material, probably as residues from the rolling of the original sheet. The obtained results suggest that despite the high laser power applied in the consecutive macro- and micro-polishing steps, the LP does not appear to remove all residues on the Nb material, but at most redistributes them on the samples. Eventually, the laser processing needs to be carried out under vacuum, so that particulates may be evaporated and removed via the pumping units (see, e.g., [21,25]).

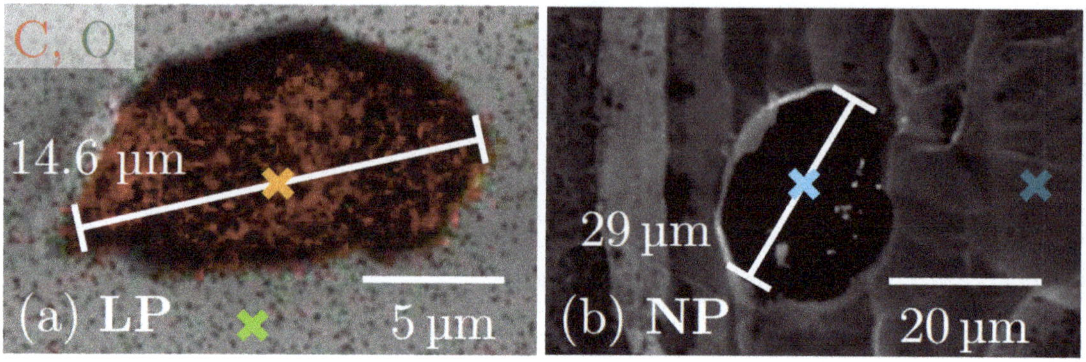


**Figure 5.** (**a**) SEM micrograph of a defect on the surface of a laser-polished Nb sample, with superimposed EDX signals of carbon (red dots) and oxygen (green dots). A spot analysis was performed for spots marked by orange and green crosses, yielding 48.0 ± 1.5 atomic % carbon and 30.8 ± 1.7 atomic % oxygen on the particle, and 16.3 ± 1.8 atomic % carbon and 9.0 ± 1.7 atomic % oxygen in the polished area. In addition, the remainder is niobium in both cases. (**b**) SEM micrograph of an untreated, pristine Nb sheet with a defect. Point analysis yields 27.9 ± 1.0 atomic % carbon and 15.4 ± 0.7 atomic % oxygen on the particle (bright blue cross), and 18.5 ± 1.1 atomic % carbon and 7.1 ± 0.5 atomic % oxygen in the pristine Nb surface area besides (dark blue cross).

In order to investigate the crystallographic structure of the laser-polished Nb materials after the different treatments, X-ray diffraction experiments have been performed, and typical results are compiled in Figure 6. For the used X-ray wavelength of 0.154 nm (Cu K$\alpha$ radiation), the penetration depth of the X-rays amounts to ca. 20 µm, so that only the near-surface region of the laser-polished samples is probed. It is important to keep in mind that the superconducting currents in a real Nb cavity only occur in a certain, even smaller depth, so that the diffraction measurements provide information about a practically relevant region of the laser-treated samples. The important parameter is the London penetration depth, which has a value of ca. 30–40 nm for niobium [38], i.e., substantially smaller compared to the depth probed by XRD. As can be seen in Figure 6a, all Bragg peaks expected for a polycrystalline body-centered cubic (bcc, space group 229) Nb sample with a lattice parameter a = 3.30 Å (ref. [36], p. 20) can be clearly detected, suggesting that the polycrystalline bcc structure persists after re-solidification. Looking in more detail at the positions of the Bragg peaks, smaller shifts are detectable, most prominently at larger Bragg angles. We thus analyzed the (400) reflection in more detail; see Figure 6b. As can be seen, the (400) peak is generally shifted towards larger Bragg angles after the different laser-polishing treatments, while the peak position of the untreated pristine Nb metal sheet corresponds well to the bcc-structure data of Nb (dashed vertical lines in Figure 6b). One may argue that the observed Bragg angle shifts arise from a possible misalignment of the diffraction geometry due to a bending of the investigated samples. However, such a bending would affect all Bragg peaks in a similar manner, which, on the other hand, is not observed.

In contrast, the observed peaks systematically shift to the theoretical position with increasing power of the micro-polishing treatment, suggesting that a compressive stress is preserved in the macro-polished Nb sheet. Such stress may result from the fast re-solidification processes of the metal in the cooling phase directly after the laser treatment. The XRD results suggest that this stress is reduced by the successive micro-polishing, where, in particular, the near-surface region of the laser-treated metal is remelted again. As can be deduced from the inset of Figure 6b, the Bragg peak shifts, and accordingly, the stress in the micro-polished samples decreases with increasing laser power during the micro-polishing treatment, again suggesting a laser power around 80–100 W as optimal.

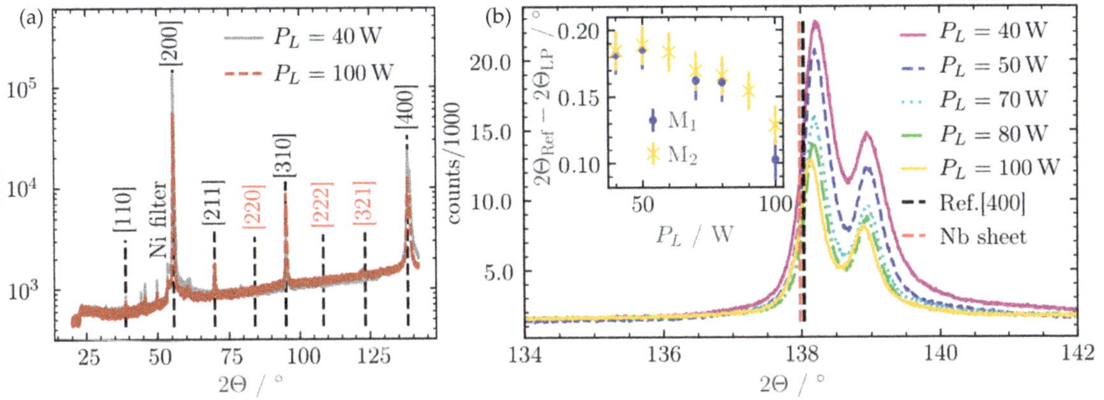


**Figure 6.** (**a**) XRD raw data for a micro-polished sample after two treatments with different laser powers. Bragg peaks are labeled with their Miller indices. Small diffraction features between 40° and 60° in the Bragg angle are due to a tungsten contamination of the Cu anode. Note the logarithmic scale of the diffractogram. (**b**) Detailed view of the (400) reflection for different micro-polishing conditions as indicated by different colored lines. In the inset, the (400) peak position is evaluated as a function of the laser power during micro-polishing for two samples, M1 and M2, with high reproducibility.

Finally, light microscopy experiments were performed on cross-sectioned samples in the pristine state and after different laser-polishing treatments on 15 × 15 mm$^2$ areas. For this purpose, a laser-polished sample was bisected, with the intersection in a region with both only macro- and macro- and micro-polished areas and untreated regions. Before microscopic inspection, the cross-sections were etched by buffered chemical etching for 5–10 min to allow a more detailed visualization with an Olympus MX40 light microscope. Some representative results are presented in Figure 7. It should be noted that the laser-polished side is pointing downwards. First of all, it is obvious that due to the stress induced by the macro-polishing treatment, the entire niobium metal sheet is slightly deformed, as indicated by the curvature of the metal in the region of the laser treatments, as visible in Figure 7b, with a maximum deviation of ca. 0.2 mm on a length of 21 mm. Since the deformation is also found in a region that is only macro-polished, the deformation can be clearly assigned to the large heat introduced in the macro-polishing step. This deformation is in qualitative agreement with the peak shifts detected in the XRD measurements, although the latter are related to the microstructure of the materials.

Measurements of the hardness were performed using Zwick ZHV 30 zwickiLine HD (https://www.zwickroell.com/, accessed on 15 october 2025) at the positions marked by red points; however, there were no meaningful changes between the untreated metal and a laser polished region. More obviously, the boundary between the remelted regions and the pristine metal can be recognized by a change in the texture in the respective regions, at a depth of about 1100 µm, as indicated by the white bars in Figure 7b. Changes are visible in the textures of the respective regions, and the most intriguing feature is a larger average crystallite size of the laser-treated regions compared to the pristine Nb sheet. Thus, the laser treatment improves the crystallinity of the metal, reducing contributions of grain boundaries to superconductivity, and thus may improve the superconducting properties accordingly.

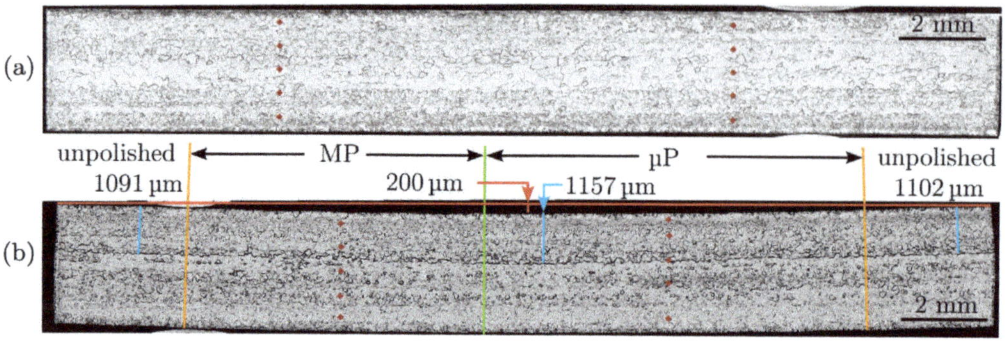


**Figure 7.** Light microscopy from a cross-section of the laser-polished niobium material (**a**) pristine Nb sheet and (**b**) laser-polished Nb sheet. Local hardness measurements were performed at the positions of the red dots, without any meaningful changes. Two polishing regions are displayed in (**b**)—one that is only macro-polished (left) and a second with macro- and micro-polishing treatments (right).

In order to verify the positive influence of the laser-polishing treatments on the parasitic field emission and the achievable acceleration fields, field emission measurements of some selected samples have been performed, i.e., the emission current was measured locally as a function of the applied electric field E; see Figure 8a. In general, as already suggested by previous experiments [23,24], a substantially larger onset field $E_{On}$ for field emission was detected for the laser-polished surfaces in comparison to the untreated Nb surface. As can be seen in Figure 8a, the pristine Nb shows an onset field for field emission of about $E_{On}$ = 90 ± 10 MV/m, at which a field emission current of 1 nA was detected. In contrast, on defect-free laser-polished surface regions, an acceleration field of almost $E_{On}$ = 800 MV/m can be applied without any indications of electron field emission, suggesting that there are no protrusions, defects, or local inhomogeneities that may facilitate electron emission (see, e.g., ref. [39]). However, without current limitation, electron field emission may be activated by large accelerating gradients, resulting in a current breakthrough with currents in the range of several µA (explosive electron emission) and related morphological changes in the inspected surfaces, as can be seen in Figure 8b. Due to the strong local electron current, the surface is heated and Nb material is emitted, leading to an irreversible deterioration of the formerly smooth surface.

For the different defect structures discussed above, substantially lower onset fields for electron field emission were detected in general: particles (Figure 5, see the SEM micrograph in Figure 8c) lead to reduced onset fields of about $E_{On}$ = 130 ± 20 MV/m; cracks (Figure 4b; see the SEM micrograph in Figure 8d) may allow electric fields of up to $E_{On}$ = 330 ± 20 MV/m prior to field emission, while the pristine Nb surface with grooves and sharp surface features reveals an onset field of about $E_{On}$ = 90 ± 10 MV/m, as already mentioned (see Figure 8e). In general, a hysteresis is observed in the current-voltage characteristics, i.e., if the electric field is reversed and scanned downwards, larger emission currents are detected (Figure 8a), which is a typical observation for activation processes triggered by the field emission. In light of the envisaged use of laser polishing treatment for a potential application to Nb cavity materials for applications in accelerators, where the accelerating field gradient should be as high as possible, it is important to note that the observed defect structures that reduce the onset field may be avoided, i.e., by improved cleaning procedures such as dry ice cleaning [10] and a second, localized laser polishing treatment [23,24], so that, thus, the use of laser polishing procedures appears very promising to suppress parasitic electron field emission processes.

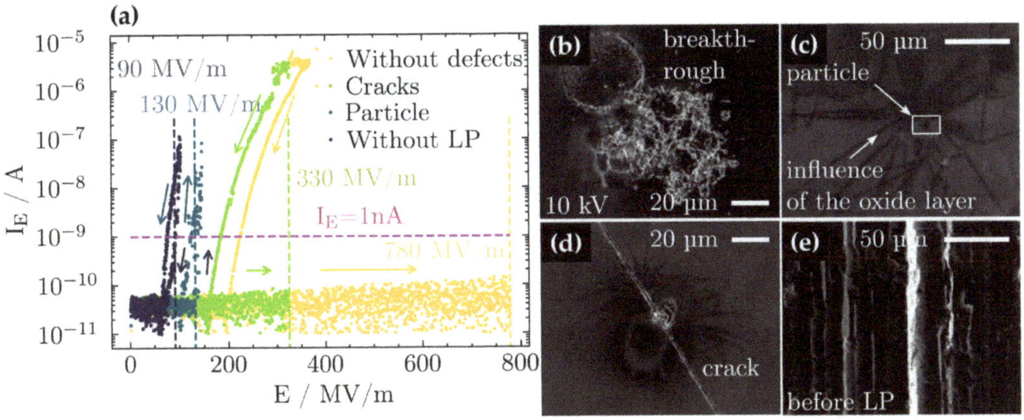


**Figure 8.** (**a**) Field emission measurements, i.e., the detected electron currents as a function of the applied electric field E in up- and downward directions as indicated by colored arrows, for different positions on the samples as indicated. The onset field was defined as the electric field for which an electron emission current of 1 nA is measured and is indicated on the respective curves. (**b–e**) SEM micrographs related to the different positions of field emission measurements are as follows: (**b**) sample position after current breakthrough at 780 MV/m (yellow curve in (**a**)), (**c**) sample position with particle on the surface (light blue curve in (**a**)), (**d**) sample position with a local crack (green curve in (**a**)), and (**e**) pristine sample (dark blue curve in (**a**)).

## 4. Conclusions

Fine-grained Nb metal sheets were successfully laser-polished, employing successive laser macro- and micro-polishing steps, and investigated by a combination of different interferometric and microscopic techniques, as well as X-ray diffraction and field emission measurements. Under process conditions close to an industrial scale, i.e., using fast-scanning laser beams in an inert noble gas atmosphere protecting the delicate Nb metal surface from oxidation, smooth surfaces with a wavy character may be prepared under optimized laser polishing conditions. These smooth surfaces allow the application of acceleration field gradients of up to almost 800 MV/m prior to the onset of parasitic electron field emission, in contrast to only 90 MV/m for the untreated niobium and 100–300 MV/m on defect-containing areas of the laser-polished samples. Furthermore, the laser-polished surfaces also reveal an improved crystallographic structure with larger grains, and even after treatment in a technical environment, no substantial oxidation.

Defects that are still detectable on the laser-polished Nb surfaces, i.e., carbon contaminations and crack formation, need further investigation; however, e.g., heat treatments as well as additional laser-polishing procedures on preheated Nb may help to reduce or even prevent these defects. The presented results are encouraging to consider laser polishing as a promising alternative to conventional polishing treatments, as huge amounts of toxic and potentially harmful acids are avoided by entirely dry laser treatments. Furthermore, laser polishing efficiently uses the applied laser energy to modify the surface and volume microstructure of the treated metal, as the laser light is efficiently absorbed by the workpieces, so that an environmentally friendly and economic processing appears feasible. Superconducting properties of laser-polished fine-grain niobium samples have not yet been measured, but are, however, foreseen for the near future, to stress that such a strategy can be promising not only for future large-scale accelerator facilities, but also for small-scale accelerators, e.g., for industrial or medical purposes.

**Supplementary Materials:** The following supporting information can be downloaded at: https://www.mdpi.com/article/10.3390/ma18215034/s1. Figure S1: Pre-screening of the parameters for the laser polishing treatments. (a) Macro polishing with a CW laser power of $P_L$(MP) = 240 W, with different line feed rates (polishing speed $v_s$(MP)). (b) Influence of the number of MP cycles with $P_L$(MP) = 240 W on the resulting roughness of the Nb sample. (c) Influence of the laser power during micro-polishing on the roughness; Figure S2: Electron-excited X-ray analysis of a laser-polished niobium sheet. Substantial amounts of carbon and oxygen are detected if an area of about 1.2 $mm^2$ is scanned. The average concentrations of niobium, carbon, and oxygen are 86.0 $\pm$ 0.4 atomic %, 4.1 $\pm$ 0.4 atomic %, and 9.9 $\pm$ 0.6 atomic % for the pristine sample, and 83.7 $\pm$ 0.4 atomic %, 7.2 $\pm$ 0.4 atomic %, and 9.1 $\pm$ 0.5 atomic % after the laser polishing. In the inset, a carbon-containing defect is shown, together with the overlaid signals of carbon (red dots) and oxygen (green dots).

**Author Contributions:** Conceptualization, F.B., L.K., E.W. and D.L.-H.; methodology, F.B., L.K., E.W. and D.L.-H.; software, F.B.; validation, F.B. and D.L.-H.; formal analysis, F.B. and L.K.; investigation, F.B. and L.K.; resources, D.L.-H.; data curation, F.B. and D.L.-H.; writing—original draft preparation, F.B. and D.L.-H.; writing—review and editing, F.B., L.K., E.W. and D.L.-H.; visualization, F.B.; supervision, E.W. and D.L.-H.; project administration, E.W. and D.L.-H.; funding acquisition, D.L.-H. All authors have read and agreed to the published version of the manuscript.

**Funding:** This project was financially supported by the German Federal Ministry of Education and Research (Grants No. 05H21PXRB1 and 05H24PXA).

**Institutional Review Board Statement:** Not applicable.

**Informed Consent Statement:** Not applicable.

**Data Availability Statement:** The original contributions presented in this study are included in the article/Supplementary Material. Further inquiries can be directed to the corresponding author.

**Acknowledgments:** We would like to thank E. Ermakova and A. Ermakov (DESY, Germany) for the preparation of the cross-sectioned samples, and the related light microscopy and hardness measurements. Fruitful discussions with M. Wenskat (DESY, Germany) are gratefully acknowledged. The research presented in this manuscript was funded by the German Federal Ministry of Education and Research under Project No. 05H21PXRB1 (TOSCA) and 05H24PXA (SuperSurfer).

**Conflicts of Interest:** The authors declare no conflicts of interest.

## References


1. Padamsee, H.; Knobloch, J.; Hays, T. *RF Superconductivity for Accelerators*; Wiley-VCH: Weinheim, Germany, 2008.
2. Darvea, C.; Elias, N.; Molloy, S.; Bosland, P.; Renard, B.; Bousson, S.; Olivier, G.; Reynet, D.; Thermeau, J.P. The ESS supercondu ing RF cavity and cryomodule cryogenic processes. *Phys. Procedia* **2015**, *67*, 1123. [CrossRef]
3. Altarelli, M.; Brinkmann, R.; Chergui, M.; Decking, W.; Dobson, B.; Düsterer, S.; Grübel, G.; Graeff, W.; Graafsma, H.; Hajd J.; et al. *XFEL Technical Design Report*. DESY 2006-097. 2006. Available online: https://xfel.desy.de/localfsExplorer_rea currentPath=/afs/desy.de/group/xfel/wof/EPT/TDR/XFEL-TDR-ExecutiveSummary.pdf (accessed on 15 October 2025).
4. Sekutowicz, J. New Geometries: Elliptical Cavities. In Proceedings of the Workshop on Pushing the Limits of RF Superconductivi Argonne, IL, USA, 22–24 September 2004; Kim, K.-J., Eyberger, C., Eds.; ANL-05/10. Argonne National Laboratory ANL: Argonn IL, USA, 2005; Volume 3.
5. Bafia, D.; Grassellino, A.; Sung, Z.; Romanenko, A.; Melnychuk, O.S. Gradients of 50 MV/m in TESLA-shaped cavities v modified low temperature bake. In Proceedings of the 19th International Conference on RF Superconductivity, Dresden, German 30 June–5 July 2019. [CrossRef]
6. Grasselino, A.; Romanenko, A.; Sergatskov, D.; Melnychuk, O.; Trenikhina, Y.; Crawford, A.; Rowe, A.; Wong, M.; Khabiboullin T.; Barkov, F. Nitrogen and argon doping of niobium for superconducting radio frequency cavities: A pathway to highly efficie accelerating structures. *Supercond. Sci. Technol.* **2013**, *26*, 102001. [CrossRef]
7. Lilje, L.; Antoine, C.; Benvenuti, C.; Bloess, D.; Charrier, J.-P.; Chiaveri, E.; Ferreira, L.; Losito, R.; Metheisen, A.; Preis, H.; et Improved surface treatment of the superconducting TESLA cavities. *Nucl. Instrum. Methods Phys. Res. A* **2004**, *516*, 213. [CrossRe
8. Navitski, A.; Prudnikava, A.; Tamashevich, A. R&D on SRF cavities at DESY towards the ILC performance goal. In Proceeding of the 5th International Particle Accelerator Conference IPAC2014, Dresden, Germany, 15–20 June 2014. WEPRI011. [CrossRef]

Santana Leitner, M.; Ge, L.; Li, Z.; Xu, C.; Adolphsen, C.; Ross, M.; Carrasco, M. Studies of radiation fields of LCLS-II super conducting radio frequency cavities. *Int. J. Mod. Phys. Conf. Ser.* **2016**, *44*, 1660209. [CrossRef]
Dangwal, A.; Müller, G.; Reschke, D.; Floettmann, K.; Singer, X. Effective removal of field-emitting sites from metallic surfaces by dry ice cleaning. *J. Appl. Phys.* **2007**, *102*, 044903. [CrossRef]
The Future Circular Collider (FCC) Conceptual Design Report. Available online: https://fcc-cdr.web.cern.ch/ (accessed on 15 September 2025).
Adolphsen, C.; Barone, M.; Barish, B.; Buesser, K.; Burrows, P.; Cawardine, J.; Clark, J.; Durand, H.M.; Dugan, G.; Elsen, E.; et al. The International Linear Collider Technical Design Report—Volume 3.II: Accelerator Baseline Design. *arXiv* **2013**, arXiv:1306.6328.
United Nations Climate Change. The Paris Agreement. Available online: https://unfccc.int/process-and-meetings/the-paris-agreement (accessed on 15 September 2025).
Temmler, A.; Willenborg, E.; Wissenbach, K. Design Surfaces by Laser Remelting. *Phys. Procedia* **2011**, *12*, 419. [CrossRef]
Zhao, L.; Klopf, J.M.; Reece, C.E.; Kelley, M.J. Laser polishing of niobium for superconducting radio-frequency accelerator applications. *Phys. Rev. Accel. Beams* **2014**, *17*, 083502. [CrossRef]
Krishnan, A.; Fang, F. Review on mechanism and process of surface polishing using lasers. *Front. Mech. Eng.* **2019**, *14*, 299. [CrossRef]
Chryssolouris, G.; Anifantis, N.; Karagiannis, S. Laser assisted machining: An overview. *J. Manuf. Sci. Eng.* **1997**, *119*, 766. [CrossRef]
Nüsser, C.; Wehrmann, I.; Willenborg, E. Influence of Intensity Distribution and Pulse Duration on Laser Micro Polishing. *Phys. Procedia* **2011**, *12*, 462. [CrossRef]
Kumstel, J.; Kirsch, B. Polishing Titanium- and Nickel-based Alloys using Cw-Laser Radiation. *Phys. Procedia* **2013**, *41*, 362. [CrossRef]
Bordatchev, E.V.; Hafiz, A.M.K.; Remus Tutunea-Fatan, O. Performance of laser polishing in finishing of metallic surfaces. *Int. J. Adv. Man. Technol.* **2014**, *73*, 35. [CrossRef]
Zhu, G.; Xu, Z.; Jin, Y.; Chen, X.; Yang, L.; Xu, J.; Shan, D.; Chen, Y.; Guo, B. Mechanism and application of laser cleaning: A review. *Opt. Laser Eng.* **2022**, *157*, 107130. [CrossRef]
Available online: https://www.cleanlaser.de/en/ (accessed on 15 September 2025).
Porshyn, V.; Serbun, P.; Bürger, H.; Soykarci, S.; Lützenkirchen-Hecht, D. Laser treatment of niobium surfaces for SRF applications. *J. Phys. Conf. Ser.* **2018**, *1067*, 082011. [CrossRef]
Porshyn, V.; Serbun, P.; Lützenkirchen-Hecht, D. Field emission from laser-processed niobium (110) single crystals. *Phys. Rev. Accel. Beams* **2019**, *22*, 023101. [CrossRef]
Brockner, F.; Lützenkirchen-Hecht, D. A versatile setup for nanosecond laser polishing processes with in situ analysis capabilities. *Rev. Sci. Instrum.* **2024**, *95*, 043003. [CrossRef]
Brockner, F.; Lützenkirchen-Hecht, D. Simultaneous laser polishing and N-doping of niobium. *Appl. Phys. Lett.* **2024**, *125*, 201902. [CrossRef]
Available online: https://www.ilt.fraunhofer.de/en/service-offers/polishing.html (accessed on 15 September 2025).
Willenborg, E. Polieren von Werkzeugstählen mit Laserstrahlung (Polishing of Tool Steel Materials with Laser Radiation). Doctoral Thesis, RWTH Aachen University, Aachen, Germany, 2006; pp. 21–23, 64–81, 135–137, ISBN 978-3-8322-4896-3.
Nüsser, C.; Kumstel, J.; Kiedrowski, T.; Diatlov, A.; Willenborg, E. Process- and Material-Induced Surface Structures During Laser Polishing. *Adv. Eng. Mater.* **2015**, *17*, 268. [CrossRef]
Nüsser, C.; Willenborg, E. High-speed laser micro polishing of $TiAl_6V_4$. *J. Laser Appl.* **2022**, *34*, 012025. [CrossRef]
Kreinest, L.; Schüssler, J.; Özaydin, O.; Kochuthundil Subhash, S.; Willenborg, E.; Bührig-Polaczek, A. Laser Remelting of Ductile Cast Iron to Achieve a Graphite-Free Surface Layer for Enabling a Manual High-Gloss Finish. *Metals* **2024**, *14*, 347. [CrossRef]
Serbun, P.; Porshyn, V.; Müller, G.; Lützenkirchen-Hecht, D. Advanced field emission measurement techniques for research on modern cold cathode materials and their applications for transmission-type x-ray sources. *Rev. Sci. Instrum.* **2020**, *91*, 083906. [CrossRef]
Elson, J.; Bennett, J. Calculation of the power spectral density from surface profile data. *Appl. Opt.* **1995**, *34*, 201. [CrossRef] [PubMed]
Gong, Y.; Misture, S.T.; Gao, P.; Mellott, P. Surface roughness measurements using power spectrum density analysis with enhanced spatial correlation length. *J. Phys. Chem. C* **2016**, *120*, 22358. [CrossRef]
Bonse, J.; Kirner, S.V.; Krüger, J. Laser-induced periodic surface structures (LIPSS). In *Handbook of Laser Micro- and Nano-Engineering*; Sugioka, K., Ed.; Springer Nature AG: Cham, Switzerland, 2021; pp. 879–936.
Kittel, C. *Introduction to Solid State Physics*, 8th ed.; John Wiley & Sons: New York, NY, USA, 2005.
Lide, D.R. *CRC Handbook of Chemistry and Physics*, 85th ed.; CRC Press: Boca Raton, FL, USA, 2004.

38. Maxfield, B.W.; McLean, W.L. Superconducting Penetration Depth of Niobium. *Phys. Rev. A* **1965**, *139*, 1515. [CrossRef]
39. Navitski, A.; Lagotzky, S.; Reschke, D.; Singer, X.; Müller, G. Field emitter activation on cleaned crystalline niobium surfac relevant for superconducting rf technology. *Phys. Rev. Spec. Top.—Accel. Beams* **2013**, *16*, 112001. [CrossRef]

MDPI AG
Grosspeteranlage 5
4052 Basel
Switzerland
Tel.: +41 61 683 77 34

*Materials* Editorial Office
E-mail: materials@mdpi.com
www.mdpi.com/journal/materials

www.ingramcontent.com/pod-product-compliance
Lightning Source LLC
LaVergne TN
LVHW070446061126
851858LV00011B/168
* 9 7 8 3 7 2 5 8 8 1 8 9 5 *